PREFACE

The 1993 ASCE National Convention technical sessions, Unsaturated Soils I: Unsaturated Soils in Engineering Practice, and Unsaturated Soils II: Mitigation of Expansive Soil Damage, were co-organized by the Soil Properties Committee, subcommittee on Unsaturated Soils, and the Shallow Foundations Committee. The goal of the two technical sessions and the associated proceedings is to share advances in the geotechnical practice and research in analysis and design for unsaturated soils.

Recent appreciation for the pervasiveness of unsaturated soils in engineering practice has led to increased interest in unsaturated soil mechanics. Traditionally, unsaturated soils have been primarily thought of as moisture-sensitive, "problem" soils such as expansive clays or collapsible silts. However, recent experience with wetting-induced swell and collapse of compacted fills has prompted many engineers to adopt a broader view of expansive and collapsible soils. Similarly, a more encompassing view of unsaturated soils, in general, has developed, and attempts toward "unification" of saturated and unsaturated soil mechanics have been made. Whether unsaturated soil mechanics should be viewed separately from traditional, saturated soil mechanics is a debatable question. However, the significance of unsaturated soils in engineering practice is clear. Unsaturated soil behavior is important to foundation design and performance, soil movements, expansive soils, collapsible soils, compacted soils, and vadose-zone dominated geo-environmental problems, to name a few engineering issues.

The papers presented in these proceedings clearly demonstrate that soil suction, and its role in the volume change, shear strength, and flow characteristics of unsaturated soils, is perhaps the most important factor in unsaturated soil mechanics. The broad-based nature of the topics included in the Unsaturated Soils sessions reflects the magnitude of unsaturated soils issues to be addressed by the geotechnical engineer.

Each of the papers included in the Proceedings has received two positive peer reviews. All papers are eligible for discussion in the *Journal of Geotechnical Engineering of ASCE*. All of the papers in these proceedings are eligible for ASCE awards.

Sandra L. Houston, M., ASCE
Associate Professor of Civil Engineering
Arizona State University

and

Warren K. Wray, M., ASCE
Professor, Department of Civil Engineering
Texas Tech University

UNSATURATED SOILS

Proceedings of sessions sponsored by the Subcommittee on Unsaturated Soils (Committee on Soil Properties) and the Committee on Shallow Foundations of the Geotechnical Engineering Division of the American Society of Civil Engineers in conjunction with the ASCE Convention in Dallas, Texas, October 24-28, 1993

Geotechnical Special Publication No. 39

Edited by Sandra L. Houston and Warren K. Wray

Published by the
American Society of Civil Engineers
345 East 47th Street
New York, NY 10017-2398

ABSTRACT

This proceedings, *Unsaturated Soils,* consists of papers presented at sessions of the 1993 ASCE National Convention and Exposition held in Dallas, Texas, October 24-28, 1993. The papers include both practice and research in the analysis and design of geotechnical engineering and geo-environmental works in unsaturated soils. The first part on unsaturated soil behavior addresses the history, as well as current trends in unsaturated soil mechanics, methods of predicting volume change of compacted and naturally- occurring soils, laboratory and field methods for obtaining unsaturated soil characteristics unsaturated flow, and unsaturated hydraulic conductivity determination. The second part on mitigation of expansive soil damage addresses engineering practice and case histories on shrink/swell behavior, performance of moisture barriers, and performance of foundations while it focuses on successful methods of avoiding or mitigating damage for soil movements. The issues addressed in these proceedings are intended to demonstrate the wide range on unsaturated soils problems and to show advances in unsaturated soil mechanics practice.

Library of Congress Cataloging-in-Publication Data

Unsaturated soils: proceedings of sessions sponsored by the
    Subcommittee on Unsaturated Soils (Committee on Soil
    Properties) and the Committee on Shallow Foundations of
    the Geotechnical Engineering Division of the American
    Society of Civil Engineers, in conjunction with the ASCE
    Convention in Dallas, Texas, October 24- 28, 1993/edited
    by Sandra L. Houston and Warren K. Wray.
        p. cm.—(Geotechnical special publication; no.39)
    Includes index.
    ISBN 0-87262-988-0
    1.Soil mechanics—Congresses. 2.Soil moisture—
Congresses. I.Houston, Sandra L. II.Wray, Warren K.
III.American Society of Civil Engineers. Geotechnical
Engineering Division. Subcommittee on Unsaturated Soils.
IV.American Society of Civil Engineers. Geotechnical
Engineering Division. Committee on Shallow Foundations.
V.ASCE National Convention (1993: Dallas, Texas) VI.Series

TA710.A1U57 1993
624.1'5136—dc20   93-31662
                                                      CIP

CONTENTS

A. Unsaturated Soils I: Unsaturated Soils in Engineering Practice

B. Unsaturated Soils II: Mitigation of Expansive Soil Damage

GEOTECHNICAL SPECIAL PUBLICATIONS

1) TERZAGHI LECTURES
2) GEOTECHNICAL ASPECTS OF STIFF AND HARD CLAYS
3) LANDSLIDE DAMS: PROCESSES RISK, AND MITIGATION
4) TIEBACKS FOR BULKHEADS
5) SETTLEMENT OF SHALLOW FOUNDATION AND COHESIONLESS SOILS: DESIGN AND PERFORMANCE
6) USE OF IN SITU TESTS IN GEOTECHNICAL ENGINEERING
7) TIMBER BULKHEADS
8) FOUNDATIONS FOR TRANSMISSION LINE TOWERS
9) FOUNDATIONS AND EXCAVATIONS IN DECOMPOSED ROCK OF THE PIEDMONT PROVINCE
10) ENGINEERING ASPECTS OF SOIL EROSION DISPERSIVE CLAYS AND LOESS
11) DYNAMIC RESPONSE OF PILE FOUNDATIONS-EXPERIMENT, ANALYSIS AND OBSERVATION
12) SOIL IMPROVEMENT-A TEN YEAR UPDATE
13) GEOTECHNICAL PRACTICE FOR SOLID WASTE DISPOSAL '87
14) GEOTECHNICAL ASPECTS OF KARST TERRAINS
15) MEASURED PERFORMANCE SHALLOW FOUNDATIONS
16) SPECIAL TOPICS IN FOUNDATIONS
17) SOIL PROPERTIES EVALUATION FROM CENTRIFUGAL MODELS
18) GEOSYNTHETICS FOR SOIL IMPROVEMENT
19) MINE INDUCED SUBSIDENCE: EFFECTS ON ENGINEERED STRUCTURES
20) EARTHQUAKE ENGINEERING & SOIL DYNAMICS (II)
21) HYDRAULIC FILL STRUCTURES
22) FOUNDATION ENGINEERING: CURRENT PRINCIPLES AND PRACTICES
23) PREDICTED AND OBSERVED AXIAL BEHAVIOR OF PILES
24) RESILIENT MODULI OF SOILS: LABORATORY CONDITIONS
25) DESIGN AND PERFORMANCE OF EARTH RETAINING STRUCTURES
26) WASTE CONTAINMENT SYSTEMS: CONSTRUCTION, REGULATION, AND PERFORMANCE
27) GEOTECHNICAL ENGINEERING CONGRESS
28) DETECTION OF AND CONSTRUCTION AT THE SOIL/ROCK INTERFACE
29) RECENT ADVANCES IN INSTRUMENTATION, DATA ACQUISITION AND TESTING IN SOIL DYNAMICS
30) GROUTING, SOIL IMPROVEMENT AND GEOSYNTHETICS
31) STABILITY AND PERFORMANCE OF SLOPES AND EMBANKMENTS II (A 25-YEAR PERSPECTIVE)
32) EMBANKMENT DAMS-JAMES L. SHERARD CONTRIBUTIONS
33) EXCAVATION AND SUPPORT FOR THE URBAN INFRASTRUCTURE
34) PILES UNDER DYNAMIC LOADS
35) GEOTECHNICAL PRACTICE IN DAM REHABILITATION
36) FLY ASH FOR SOIL IMPROVEMENT
37) ADVANCES IN SITE CHARACTERIZATION: DATA ACQUISITION, DATA MANAGEMENT AND DATA INTERPRETATION
38) DESIGN AND PERFORMANCE OF DEEP FOUNDATIONS: PILES AND PIERS IN SOIL AND SOFT ROCK
39) UNSATURATED SOILS

An Overview of Unsaturated Soil Behaviour

Delwyn G. Fredlund,
Professor, Department of Civil Engineering,
University of Saskatchewan, Saskatoon, Saskatchewan, Canada, S7N 0W0

Harianto Rahardjo,
Senior Lecturer, School of Civil and Structural Engineering,
Nanyang Technological University, Singapore, 2263

ABSTRACT

Traditional soil mechanics practice has experienced significant changes during the past few decades. Some of these changes are related to increased attention being given to the unsaturated soil zone above the groundwater table. Increased concerns over the environment have, in part, "fuelled" the need to better understand the behaviour of the zone near ground surface. The computational capability available to the geotechnical engineer has also strongly influenced the engineers ability to address these complex problems.

The portion of the soil profile where the pore-water pressures are negative, is known as the vadose zone. The ground surface is subjected to a flux type boundary condition for many of the problems faced by geotechnical engineers. Unsaturated soil mechanics has become a necessary tool for analysing the behaviour of soils in the vadose zone and the flux boundary conditions as required in many geotechnical and geo-environmental problems.

This paper presents the scope and nature of typical unsaturated soils problems. The basic physical relationships associated with unsaturated soil mechanics are presented. The research needs and a possible future direction for unsaturated soil mechanics are outlined. One of the conclusions is that the Soil Water Characteristic Curve can be used to estimate relevant unsaturated soil properties for engineering analyses.

INTRODUCTION

Geotechnical engineering technology and practice developed primarily in the temperate climates of the world. As a result, research was directed towards problems involving soils with positive pore-water pressures. Classic areas of study were associated with seepage, shear strength and volume change. The practice of geotechnical engineering has however, undergone continual change.

Only about six decades ago, soil mechanics moved from being primarily an art to taking on a science basis. It is only in the last four to five decades that soil mechanics has become a mandatory part of civil engineering curriculums at universities. It is only the last two to three decades that computers have greatly influenced our ability to model complex geotechnical problems. There has been continual change and now geotechnical engineering has expanded to embrace a large number of problems occurring in the geo-environmental area.

The drier climatic regions of the world have become increasingly aware of the uniqueness of their soil mechanics problems. Throughout the development of geotechnical engineering as a science, there has been an increasing interest in developing a science basis for the broad category of problems dealing with unsaturated soils. Now, a science appears to be immerging which is receiving acceptance on a global scale. This paper will attempt to provide an overview of the scope and nature of unsaturated soil behaviour.

NEED FOR AN UNSATURATED SOIL FRAMEWORK

Geotechnical engineering has traditionally been viewed as an engineering field which is strongly rooted in engineering mechanics, directed at solving problems related to strength, strain and seepage. The effective stress principle is the key concept that has led to the rapid transfer of geotechnology around the world. Similarly, the stress state variable approach is becoming the means for transferring unsaturated soil behaviour from one continent to another.

Arid and semi-arid regions of the world comprise more than one-third of the earth's surface. Soils in these regions are dry and desiccated near the ground surface. These conditions may extend to a considerable depth and in some cases the water-table may be more than 30 metres below ground surface. Even under humid climatic conditions the groundwater table can be well below the ground surface and the soils used in construction are unsaturated. Engineering organizations have realized the need for a technical discipline which more specifically addresses the soil mechanics problems related to unsaturated soils. This has brought about the formation of a committee on Soils in Arid Regions within the International Society for Soil Mechanics and Foundation Engineering (ISSMFE), the committee on Unsaturated Soils within the American Society of Civil Engineering (ASCE), and the committee on Soils in Arid Regions within the Transportation Research Board (TRB). The formation of these committees bears witness of the growing awareness of the need to better understand unsaturated soil behaviour.

Compacted soils comprise a large part of the earth structures designed by engineers. Examples are the roads, airfields, earth dams and the many other structures which form part of the infra-structure of our society. Swelling clays, collapsing soils and residual soils are all examples of unsaturated soils encountered in engineering practice. These soils are often referred to as "Problematic Soils". Common to all of these soils is their negative pore-water pressures which play an important role in their mechanical behaviour and also make them difficult to test in the laboratory.

In a matter of only about two decades, world attention shifted from the analysis of engineered structures to limiting the impacts of technology and developments on the natural world. Geotechnical engineers found themselves well positioned, by virtue of their training and experience, to study the impact of a wide range of developments. Geotechnical engineers have in general, quickly broadened the scope of their domain of practice. In North America and other parts of the world, many geotechnical consulting firms now find that more than 75% of their work involves geo-environmental type projects. As part of geo-environmental type projects, it has become necessary to study the mass flux of contaminants transported to the groundwater system. Most of these problems occur near ground surface and as such a knowledge of unsaturated soil behaviour is valuable.

HISTORICAL PERSPECTUS ON CLASSICAL SOIL MECHANICS

The effective stress variable, $(\sigma - u_w)$, became pivotal in the 1930's as a means of communicating the behaviour of saturated soils (Terzaghi, 1936). Soil mechanics moved from an empirical basis to a science basis and enjoyed the implied status. Soil behaviour was related to effective stress which was independent of the soil properties. The scope of saturated soil mechanics embraced three primary areas:

(1)   Seepage analyses where the problems were classified as either confined or unconfined flow analysis (Casagrande, 1936),

(2)   Plasticity and limit equilibrium analyses where the problems ranged from slope stability, to bearing capacity and lateral earth pressures, and

(3)   Volume change analyses directed primarily towards the prediction of settlement in soft clays.

The basic formulations in classical soil mechanics were primarily of a static and steady state nature. One exception was the theory of consolidation which illustrated the interaction of deformation and seepage. The theory of consolidation provided an excellent mathematical and rheological tool to assist the engineer in visualizing saturated soil behaviour. It allowed the prediction of pore-water pressures in time and space, and became the symbol of classical soil mechanics.

With time, there has been a need to consider a wider variety of unsteady state or transient related analyses. This has been most visible in the geo-environmental area where the geotechnical engineer is called upon to predict chemical concentrations with respect to time and space. The diffusion of a chemical is superimposed on the conductive movement in the water phase. In other words, an additional unsteady state analysis has been added to the seepage of water through a soil. The soil property, coefficient of permeability, has often been regarded as one of the most difficult soil properties to evaluate. Now, this property is the central focus of many analyses and much research has been directed towards its quantification. To complicate matters further, many of the processes of concern occur in the upper portion of the soil profile; in the vadose zone where the pore-water pressures are negative. In this zone, the coefficient of permeability is a function of the negative pore-water pressure and this results in nonlinear flow formulations. In addition, the groundwater table is no longer the upper boundary of concern. Rather, the ground surface geometry becomes the boundary for the problem being analyzed and movements through the unsaturated zone become of vital interest.

At the same time as the geotechnical problems have become more complex, the capacity to handle these problems has also improved. The microcomputer has brought a wide range of data management and analytical skills to the desk of engineers. The computer programs used on a routine basis may range from spreadsheet software to powerful coupled, transient applications software. The manner in which the applications type software is being used in a consulting firm has also taken on a new direction. The software is most often used in a parametric or sensitivity manner in order to embrace possible field conditions and thereby give the geotechnical engineer an indication of the bounds on soil behaviour.

KEY STEPS TOWARDS THE DEVELOPMENT OF UNSATURATED SOIL MECHANICS

The development of unsaturated soil mechanics has been relatively slow in comparison to saturated soil mechanics. It is interesting to note that there were a significant number of research papers related to unsaturated soil behaviour at the first International Conference on Soil Mechanics and Foundation Engineering, (ISSMFE), Harvard, in 1936. Obviously, problems related to unsaturated soils behaviour were considered to be an important part of engineering practice at that time. Most emphasis was placed on the flow of water in the unsaturated, capillary zone. A typical problem involved the study of the "syphon" effect or the effect of water flow through a dam, above the phreatic line.

Research conferences subsequent to the first ISSMFE conference, show a decreasing interest in unsaturated soil behaviour. This was no doubt, influenced by the fact that it was easier and more fruitful to direct research efforts towards saturated soil behaviour. In addition, the empirical handling of many seepage problems as unconfined flow tended to eliminate the need for consideration of the zone where the pore-water pressures were negative (i.e., the vadose zone). It was easy to rationalize away the need to devote time for the study of the vadose zone.

In the 1950's, research at Imperial College, London, was directed towards a fundamental understanding of unsaturated soil behaviour within the classical framework of saturated soils (Bishop, Alpan, Blight and Donald, 1960). The research involved careful laboratory studies, primarily on the shear strength of unsaturated soils. These studies appear to have provided the catalyst for further studies in various countries of the world.

With time, there was increasing awareness that many geotechnical problems could better be analyzed with an understanding of unsaturated soil behaviour (Fredlund and Rahardjo, 1985 and 1987). Specialty conferences were organized to exchange information on soils labelled as "problematic soils". The first such series of conferences was organized on swelling and shrinking soils and has become known as the expansive soils series of conferences. To-date, a total of 7 conferences have been held in various countries of the world. Other "problematic soils" such as residual and collapsing soils, have also been the theme of international conferences.

The behaviour of compacted soils has also been studied with the realization that they have not enjoyed the same favourable theoretical context as have saturated soils. Common to all these "problematic soils" has been the state of stress in the pore-water phase. The pore-water pressures are negative and it is a change in the pore-water pressure that produces behaviour which has been difficult to predict. Negative pore-water pressures have always proven to be difficult to measure. The difficulty in predicting the behaviour of these soils can, in part, be related to the manner in which the laboratory tests are conducted. It is often considered as acceptable practice to immerse the specimen in water prior to testing. This procedure dramatically alters the properties which need to be measured. In other words, our conventional laboratory equipment is not capable of running laboratory tests under controlled negative pore-water pressure conditions.

The basic theories associated with most of the classic areas of unsaturated soil behaviour were assembled in the 1970's and later (Fredlund, 1979). The primary deterrent to their application has been related to difficulties associated with measuring negative pore-water pressures insitu (i.e., matric suction) (Fredlund and Rahardjo, 1988). While there is still much research to be done on the measurement of matric suction, the theories of unsaturated soil behaviour are experiencing rapidly increasing acceptance. The questions which need to be answered with respect to unsaturated soils are similar in many cases to those which have been addressed for saturated soils.

The greatest impetus for the application of saturated/unsaturated seepage analysis has come about as a result of developments in the geo-environmental area. Procedures for the characterization of the permeability function for unsaturated soils have been established and the use of saturated/unsaturated computer models has become a common practice.

The formulations associated with unsaturated soils are often nonlinear and as a result their solution is computationally intensive (Papagiannakis and Fredlund, 1984). In general, this has not proven to be a deterrent due to the availability and capability of microcomputers.

THE VADOSE ZONE

The portion of the soil profile above the groundwater table is called the vadose zone (Fig. 1). This zone can be broadly subdivided into a portion immediately above the water table which remains saturated (i.e., degree of saturation is approximately 100%) even though the pore-water pressures are negative, and a portion where the soil becomes unsaturated (i.e., degree of saturation is less than 100%). The desaturation in the upper portion may be due to exceeding the air entry value of the intact soil or due to the formation of fissures and cracks.

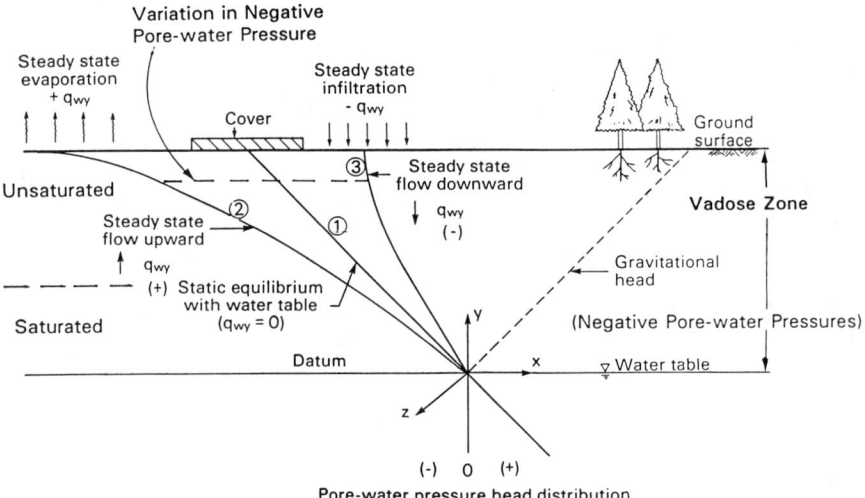

Fig. 1.   Illustration of possible negative pore-water profiles in the vadose zone

Regardless of the degree of saturation of the soil, the pore-water pressure profile will come to equilibrium at a hydrostatic condition when there is zero flux from the ground surface. If moisture is extracted from the ground surface (e.g., evaporation), the pore-water pressure profile will be drawn to the left. If moisture enters at the ground surface (e.g., infiltration), the pore-water pressure profile will be drawn to the right.

The precipitation conditions at a site are often known from past records and are available for design purposes. The evaporative flux, on the other hand, must be computed through use of one of several models which have been suggested by various researchers. Some of the most approximate estimates for geotechnical problems have been associated with the assessment of evaporative flux. Only recently have engineers began to include the role of soil suction in computing the rate of evaporation from the ground surface (Wilson, 1990). Figure 2 shows that the rate of release of water to the atmosphere, from a sand, silt or clay, is related to the suction in the soil. The computation of the evaporative flux has become an area of research which is of great value to engineers dealing with geo-environmental problems.

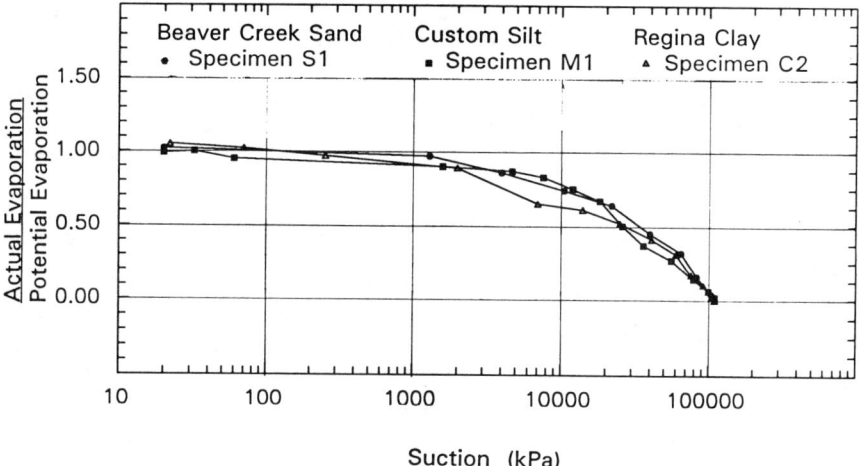

Fig. 2.   Relationship between the ratio of actual evaporation to potential evaporation, and soil suction for three soils (from Wilson, 1991).

In order for surface water to reach the groundwater table, it must pass through the vadose zone. An understanding of the behaviour of the vadose zone is important in making long-term predictions on the movement of water and subsequently, the movement of contaminants. Much effort has gone into the numerical modelling of the vadose zone. This research has been done primarily by soil scientists while geotechnical engineers have paid little attention to this topic.

There are many complexities associated with the vadose zone because of its fissured and fractured nature. The tendency in geotechnical engineering has been to avoid the analysis of this zone, if possible. However, in many cases it is an understanding of this zone which holds the key to the performance of an engineered structure.

One of the characteristics of the upper portion of the vadose zone is its ability to slowly release water vapour to the atmosphere at a rate dependent upon the permeability of the intact portions of soil. At the same time, the inflow of water can occur through the fissures under a gradient of unity. There appears to be no impedance to the inflow of water until the soil swells and the mass becomes intact, or until the fissures and cracks are filled with water.

A common misconception is that the water can always enter the soil at the ground surface. If the soil is intact, the maximum flux of water at the ground surface is the saturated coefficient of permeability of the soil. This value may be extremely low. If the ground surface is sloping, the surface layer can become saturated and have a higher coefficient of permeability than the underlying soil. As a result, water runs down the top layer of soil on the slope, and never enters the underlying soil.

QUESTIONS ASKED OF THE GEOTECHNICAL ENGINEER

The types of questions asked of engineers pertaining to the behaviour of unsaturated soils, are similar to questions addressed in saturated soil mechanics. However, questions related to unsaturated soil behaviour have not been addressed with the same degree of confidence as those for saturated soils.

When the soils involved are unsaturated, the engineer has often used the unsaturated condition as an excuse for not getting involved or else not carrying out a significant investigation or analysis. An example is the case of temporary support which must be provided to an open trench or a temporary excavation for the installation of a foundation. In this case, the problem is often left to be solved by the contractor. And one should note that contractors are, in general, quite capable of solving problems in a wise manner. The engineer should be ready to learn from these contractors, and as well add insight by way of theory and analysis in order to come up with a superior, reliable design.

Soil investigation reports often contain statements to the effect that the excavation should be able to stand in an unsupported manner for a certain period of time. In effect, the engineer is "gambling" on there being a minimum amount of rainfall while the excavation is open. The only variable which appears to be changing with time is the negative pore-water pressure in the soil in the backslope. The pore-water pressures will remain essentially constant, and the slope remains stable, unless there is an influx of water at the ground surface.

CATEGORIZATIONS FOR UNSATURATED SOILS

Classical soil mechanics can broadly be categorized as shown in Fig. 3. This classification serves to illustrate the distinctness of unsaturated soil mechanics. At the same time, the theoretical framework for unsaturated soil mechanics shows that saturated soils can be treated as a special case of the more general theories for unsaturated soils. The terminology, saturated soil mechanics and unsaturated soil mechanics, would suggest that it is the degree of saturation which provides the distinction between these two categories. While this is true, it is difficult to measure the degree of saturation and to determine a value which can be used as a dividing

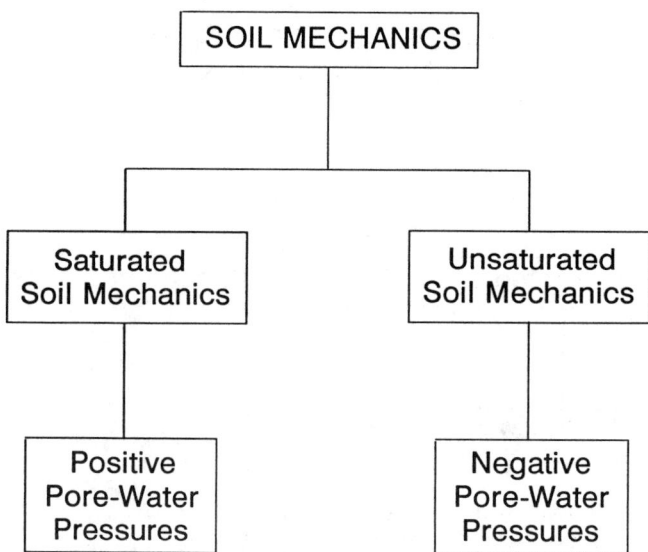

Fig. 3.   General categorization of Soil Mechanics based on the state of stress in the pore-water

line. It is rather the state of stress in the water phase which has become the indicator for classification purposes. Any soil with a negative pore-water pressure is considered to be part of unsaturated soil mechanics. It is also recognized that soils with negative pore-water pressures can be saturated or contain air bubbles in an occludded form. Figure 4 shows a soil classification according to the nature of the air and water phases.

The smallest amount of air in a soil theoretically renders the soil unsaturated. The corresponding degree of saturation will be high and the air bubbles will be in an occludded form. Soils near to the ground surface, or near to the water table, may have occludded air bubbles. Soils compacted above their optimum water content fall within this category. The occludded air bubbles render the pore fluid compressible. Little research has been directed towards problems in this category.

The physics and engineering principles involved with dry soils are essentially the same as those involved with saturated soils. The difference is that the incompressible water in the voids of a saturated soil is now replaced with a compressible pore fluid containing air bubbles.

Most attention in research has been given to the case where air and water are continuous throughout the voids. This is also the primary emphasis of this paper. It appears to be the case most relevant to engineering practice.

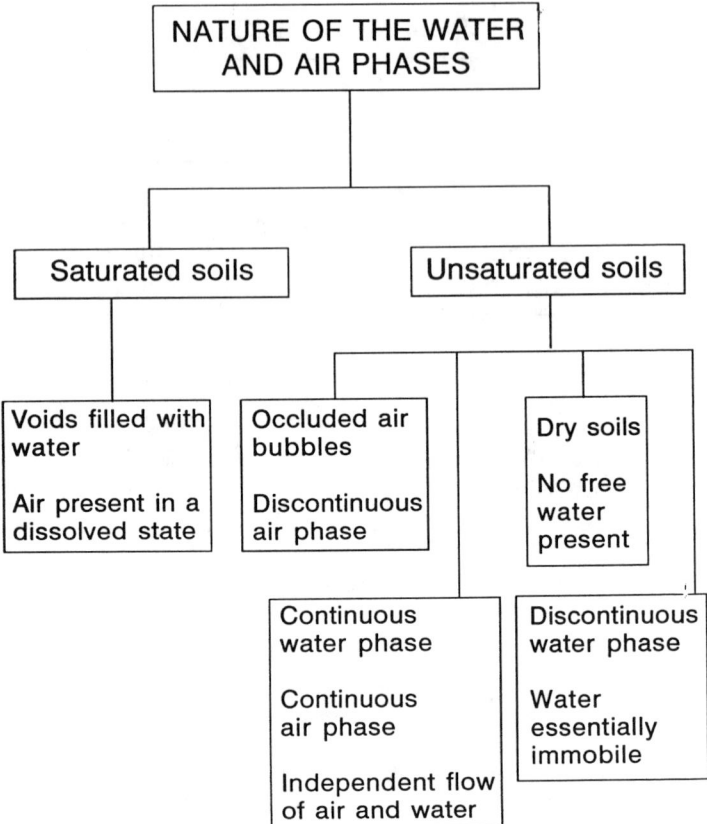

Fig. 4.   Categorization of Unsaturated Soil Mechanics based upon the degree
               of saturation

Unsaturated soil mechanics can also be categorized in a manner consistent with the
traditional classic areas of soil mechanics. There are three main categories; namely,
seepage, shear strength and volume change. This subdivision, along with examples
associated with each category, are shown in Fig. 5. This categorization shows that
the same types of problems are of interest for both saturated and unsaturated soils.

The genetic categorization of unsaturated soils has influenced the types of research
conferences which have been organized. In other words, the geologic genesis of a
deposit has certain characteristics which indicate the nature of engineering properties
likely to be encountered. Figure 6 shows a number of the common genetic categories.
From the standpoint of a theoretical framework, this categorization has little to offer
in describing the behavior of soils. In other words, essentially the same theories can

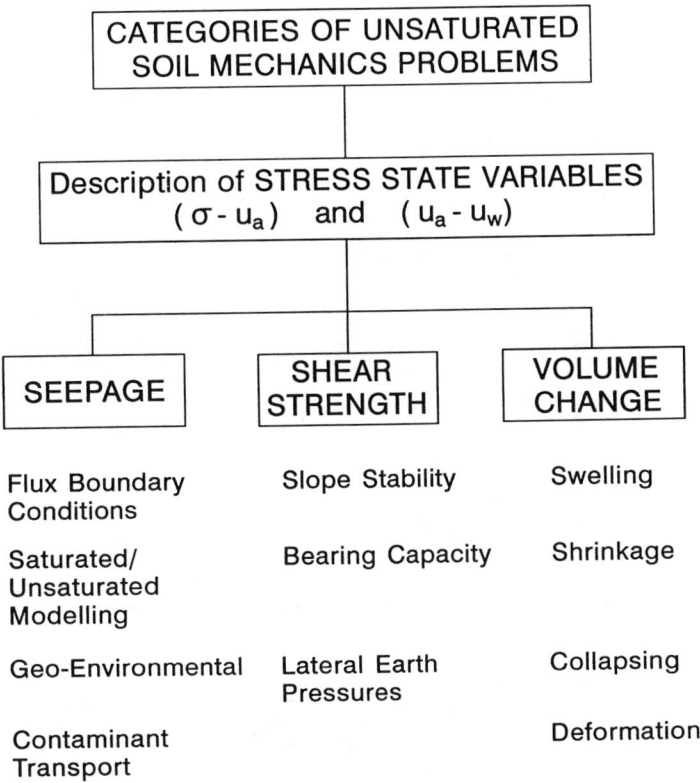

Fig. 5. Categorization of Unsaturated Soil Mechanics based on the type of engineering problem

be shown to apply for all types of soil. Probably the most meaningful part of the categorization in Fig. 6 is the breakdown between natural and compacted (or remoulded) soils. In practice, the engineer must decide which soils can be combined and treated as the same soil. Even when soils have the same genetic background, it may still be necessary to classify them as different soils because of their differing stress histories. Likewise, each compacted soil must be classified as a different soil when they are compacted at different water contents or compactive efforts (Fredlund and Rahardjo, 1993).

Another categorization of unsaturated soil mechanics is related to the volume change behaviour of the soil (Figure 7). A soil may either increase or decrease in overall volume when subjected to wetting (i.e., a decrease in matric suction). If the volume increases upon wetting, the soil has a swelling nature and if the volume decreases upon wetting, the soil has a collapsing nature. The difference in behaviour is

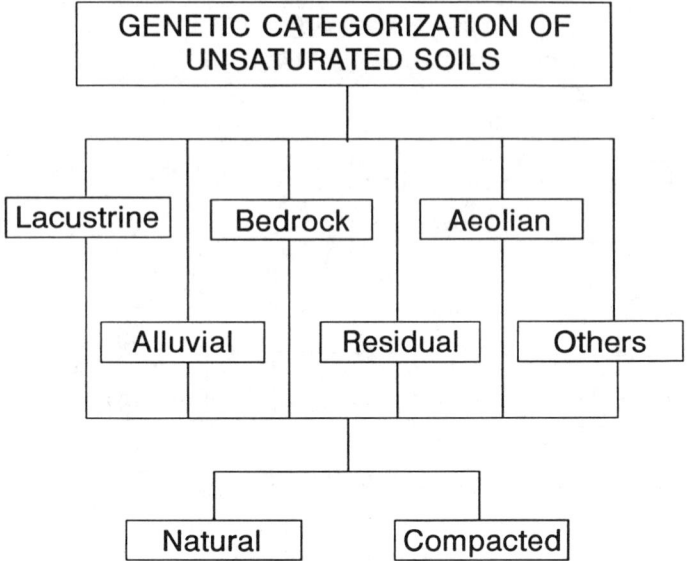

Fig. 6.   Genetic categorization of Unsaturated Soil Mechanics

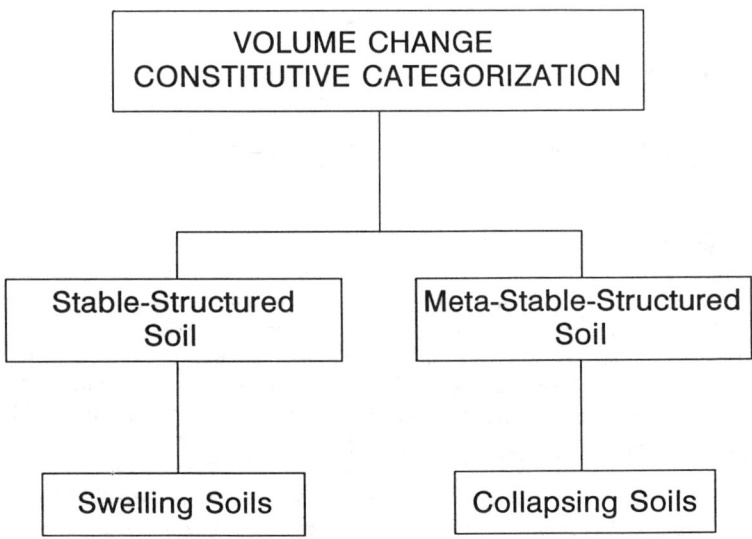

Fig. 7.   Categorization of Unsaturated Soil Mechanics based on the soil structure.

associated with the structure of the soil. The swelling soil has a "stable-structure" and the collapsing soil has a "meta-stable-structure". This categorization also applies to both remoulded (i.e., compacted) and natural soils. The awareness of these types of behaviour is important since the constitutive relationships (i.e., volume change behaviour in particular), are different for each case. The primary or basic constitutive behaviour of the soil should be defined for the case of a stable-structured soil.

There are also textural and plasticity categories of soils (i.e., sands, silts and clays). Regardless of the various soil categorizations, similar engineering problems can be encountered.

TERMINOLOGY FOR UNSATURATED SOIL MECHANICS

There has been a gradual evolution of an acceptable terminology for unsaturated soils. The terms, partly saturated and partially saturated, have, for the most part been replaced by the term, unsaturated soils. The term, unsaturated soils, is usually used in reference to all soils where the pore-water pressures are negative.

Because of the singularity of the effective stress variable for saturated soils, it has become necessary to adopt a more general description for the state variables for an unsaturated soil. The terminology associated with state variables (i.e., stress and deformation) is consistent with that used in continuum mechanics (Fung, 1965). The terminology allows for more than one state variable in the description of the stress state of the soil. As well, a clear distinction is maintained between the description of the state variables and the formulation of constitutive relations for the soil. The state variables are maintained as variables independent of the properties of the soil while the constitutive relations incorporate measureable soil properties into one-to-one form of equations.

One of the stress state variables for an unsaturated soil is matric suction, which is defined as the difference between the pore-air and the pore-water pressures. There appears to have been some confusion in the literature over the spelling of the term, matric suction. The 1963 report of the International Society of Soil Science to establish Soil Physics terminology, defined the term and spelled it as matric suction. In 1964, the Symposium-in-Print entitled, "Moisture Equilibria and Moisture Changes in Soils Beneath Covered Areas", edited by G. D. Aitchison, was presented to the First International Conference on Expansive Soils (1965). The first article in this publication was a statement of the Review Panel which sets forth the Notation and Definition of Terms for expansive soils. The proposed notations are referenced to the International Society of Soil Science. Unfortunately, the capillary suction component of the soil was spelled as matrix suction while the definition was the same as that given in 1963 by the International Society of Soil Science. This has subsequently led to confusion in the literature regarding the spelling of the capillary component of suction.

In 1970, the International Society of Soil Science Society published a Glossary of Soil

Science terms using the spelling, matric suction. In 1976, the Canada Department of Agriculture, Ottawa, published a Glossary of Terms in Soil Science, using the spelling, matric potential. In 1979, the International Society of Soil Science Society again reviewed their terminology, maintaining their previous spelling as matric. The International Society for Soil Mechanics and Foundation Engineering has never adopted terminology for the suction components of an unsaturated soil.

The research literature appears to have consistently used the spelling, matric suction, with the exception of a portion of the geotechnical literature which has used the spelling, matrix suction. It is suggested that a spelling consistent with the soil physics literature be used throughout the geotechnical community.

In general, there appears to have been an attempt to maintain the classic soil mechanics terminology for describing the behaviour of unsaturated soils. This is also true for recent extensions using stress path terminology and critical state concepts (Wheeler, 1992).

THEORETICAL FRAMEWORK FOR UNSATURATED SOIL MECHANICS

Before identifying relevant unsaturated soil properties, it is important to define the variables which can be used to represent the stress state of the soil. These are the variables which will allow engineering experience from various parts of the world to be readily compared. It will be the basis on which a transferable science can be built. The stress state is the single element which establishes geotechnical engineering as a science, set apart from an empirical art. As a result, it becomes the basis or the foundation block on which engineers build their engineering formulations.

(1)   Stress State Variables

There now appears to be a fairly general consensus that the stress state for an unsaturated soil should be described using two independent stress tensors with two independent normal stress variables (Fredlund and Morgenstern, 1977). These two variables become a logical extension of the effective stress variable used for a saturated soil and are shown in Eq. 1.

The $(\sigma - u_a)$ term is called the net normal stress and the $(u_a - u_w)$ term is called the matric suction. Other variables are defined under the Definition of Soil Terms. Figure 8 illustrates how two independent normal stresses can be used to define the stress state at a point.

The simplest way to visualize the need for two independent stress state variables is to realize that total stress changes and pore-water pressure changes do not produce equivalent responses in an unsaturated soil. This sets their behaviour apart from that of saturated soils. This is true for both shear strength and volume change behaviour.

$$\begin{bmatrix} (\sigma_x - u_a) & \tau_{yx} & \tau_{zx} \\ \tau_{xy} & (\sigma_y - u_a) & \tau_{zy} \\ \tau_{xz} & \tau_{yz} & (\sigma_z - u_a) \end{bmatrix}$$

and   [1]

$$\begin{bmatrix} (u_a - u_w) & 0 & 0 \\ 0 & (u_a - u_w) & 0 \\ 0 & 0 & (u_a - u_w) \end{bmatrix}$$

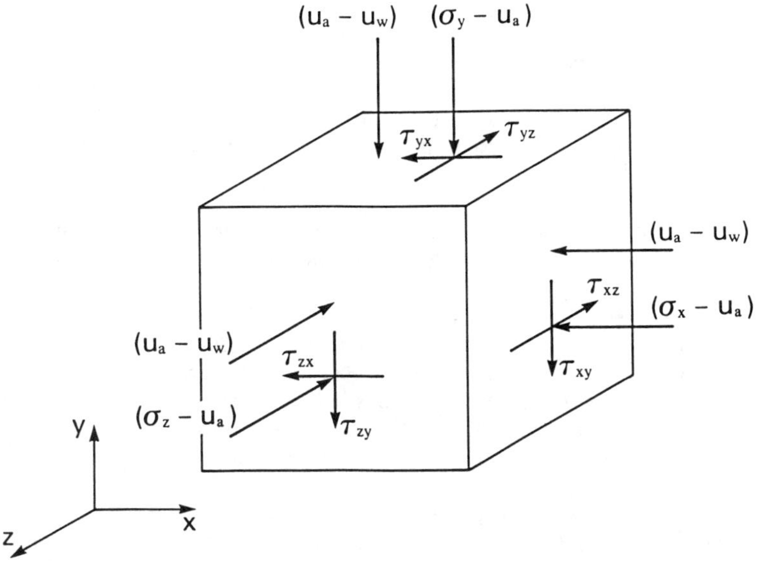

Fig. 8.   Element showing the stress state at a point in an unsaturated soil.

Since the pore-air pressure is constant (i.e., atmospheric) for most practical geotechnical problems, it is logical to independently reference the total stress and the pore-water pressure to the pore-air pressure. All descriptions of soil behaviour (i.e., volume change and shear strength) can now be referenced to the stress state variables. There is also a smooth transition from the unsaturated case to the saturated case in that the pore-air pressure becomes equal to the pore-water pressure as the soil becomes saturated.

(2)   Volume Change Moduli

The independent stress state variables can be used to formulate the constitutive relations for an unsaturated soil. The volume change constitutive equation for isotropic loading can be written in terms of a change in void ratio.

$$de = a_t d(\sigma_c - u_a) + a_m d(u_a - u_w)$$ [2]

More than one constitutive equation is required for the complete volume-mass characterization of an unsaturated soil since water content (or degree of saturation) change is independent of the void ratio change. The water content constitutive equation has generally been used as the second constitutive relationship for an unsaturated soil.

$$dw = b_t d(\sigma_c - u_a) + b_m d(u_a - u_w)$$ [3]

Equations 2 and 3 are written for the case of isotropic loading; however, similar equations can be written for other loading conditions. Figure 9 shows a graphical, three-dimensional representation of the void ratio and water content constitutive surfaces. When the matric suction of the soil is zero, the change in void ratio and the change in water content are equivalent in terms of their response to a change in total stress.

The void ratio constitutive surfaces for the loading and unloading of a stable-structured soil are shown in Fig. 10. The void ratio constitutive surface for the loading and unloading of a meta-stable-structured soil are shown in Fig. 11. The volume change behaviour differs primarily in that both a decrease or an increase in matric suction produces a decrease in volume of the soil. As a result, a ridge line appears to form across the constitutive surface. More research is needed in order to completely define the character of the constitutive surface for unsaturated soils. At the same time, it is noted that the soils used in several research programs to study constitutive behaviour have had meta-stable soil structures.

It is the water content versus matric suction relationship which becomes an important, additional relationship in quantifying unsaturated soil behaviour. Equipment for the quantification of this relationship has been developed in the soil science discipline and is known as Pressure Plate equipment (Figure 12). The data gives the relationship between water content and matric suction and is known as the "Soil Water Characteristic curve".

The Soil Water Characteristic Curve is immerging as being of great value in quantifying unsaturated soil behaviour. In general, it is possible to estimate most unsaturated soil properties from saturated soil parameters and the Soil Water Characteristic curve. The methods for approximating the unsaturated soil properties are presently an area of needed research. The total pressure has been found to influence the shape of the Soil Water Characteristic curve and it may be necessary to develop new pressure plate test apparatuses which are more suitable for geotechnical testing purposes.

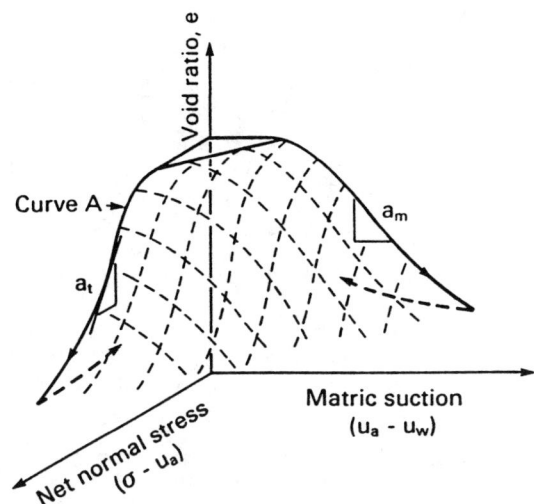

(a) Void ratio constitutive surface

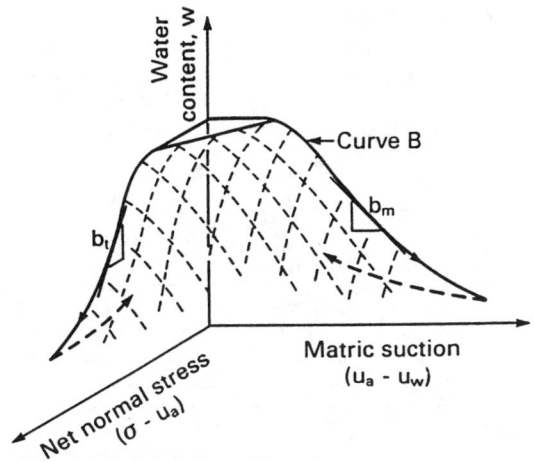

(b) Water content constitutive surface

Fig. 9.   Void ratio and water content constitutive surfaces for an unsaturated soil.

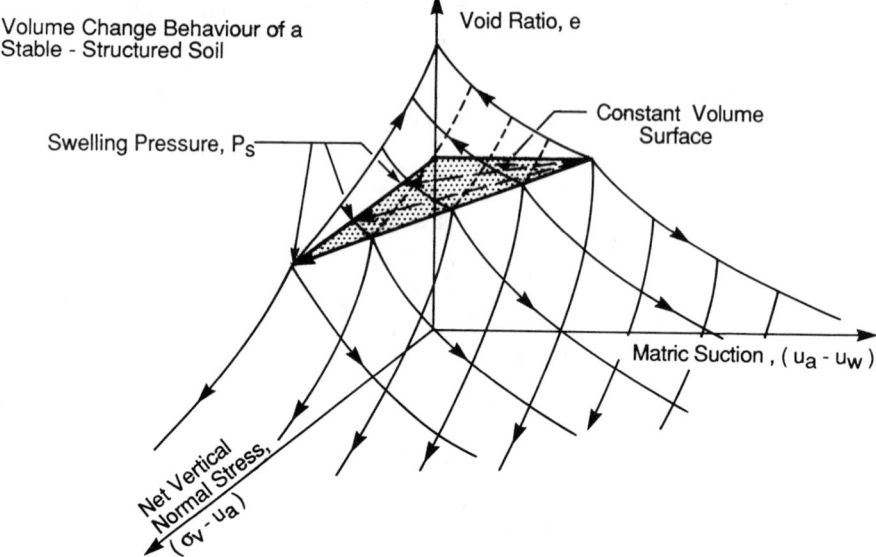

Fig. 10.   Loading and unloading constitutive surfaces for a stable-structured, unsaturated soil.

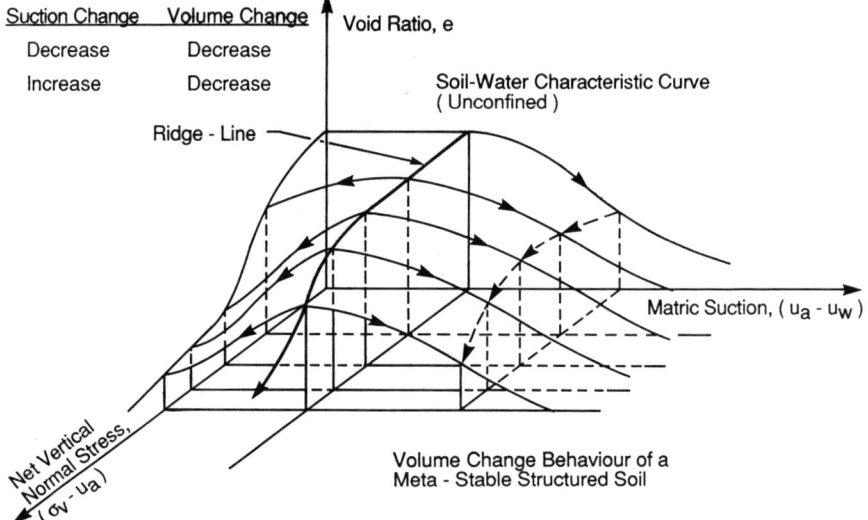

Fig. 11.   Loading and unloading void ratio constitutive surfaces for a meta-stable-structured, unsaturated soil.

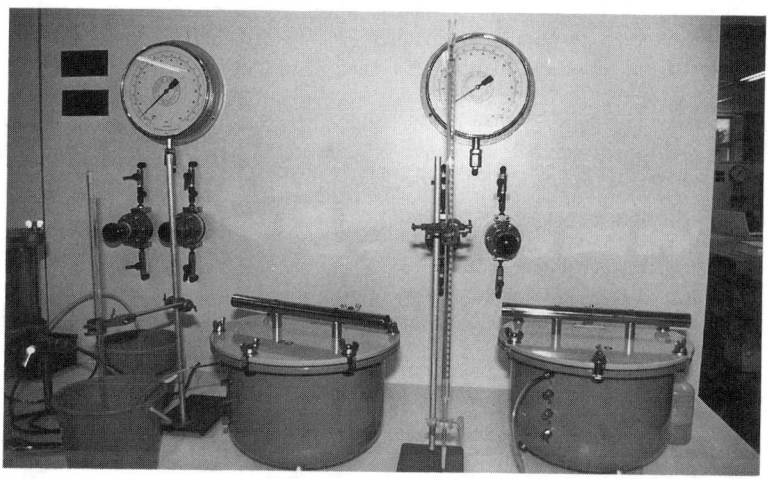

Fig. 12.   A 5-Bar Pressure Plate apparatus for measuring the Soil Water
              Characteristic Curve.

Research in soil science has led to the proposal of several mathematical
approximations for the Soil Water Characteristic Curve. The proposed equations take
the form of a cummulative frequency distribution curve and can be shown to be
special cases of the following form.

$$\theta_w = \{ 1 / (1 + a \mid u_a - u_w \mid^n) \}^m \qquad\qquad [4]$$

where: θ_w = the volumetric water content of the soil
              a = a soil parameter related to the air entry value or "bubbling
                   pressure"
              n and m = soil parameters related to the rate of water
                   extraction from the voids.

The m parameter has been set to 1.0 for many studies and this gives rise to the well-
known form advocated by Gardner (1958). Burdine, in 1953 proposed that the m and
n parameters be related as follows.

$$m = 1 - 2 / n \qquad\qquad [5]$$

Mualem (1976) suggested the following relationship between the m and n parameters.

$$m = 1 - 1 / n \qquad\qquad [6]$$

van Genuchten (1980) suggested that the relationship between the m and n values not be fixed. Further study should be undertaken to determine the most satisfactory mathematical form for the Soil Water Characteristic Curve.

It appears that approximate estimates of the unsaturated soil properties may be possible based upon the nature of the Soil Water Characteristic Curve. These properties may be satisfactory for geotechnical practice.

(3) Shear Strength Parameters:

The shear strength equation for unsaturated soils has been formulated as a linear combination of the stress state variables incorporating shear strength parameters.

$$\tau = c' + (\sigma_n - u_a) \tan \phi' + (u_a - u_w) \tan \phi^b \qquad [5]$$

Figure 13 shows a three-dimensional representation of Eq. 5. As the testing of unsaturated soils has been extended over an ever increasing suction range, and for many soil types, there has become increasing evidence that the shear strength relationship involving suction can be nonlinear (Fredlund, Rahardjo and Gan, 1987).

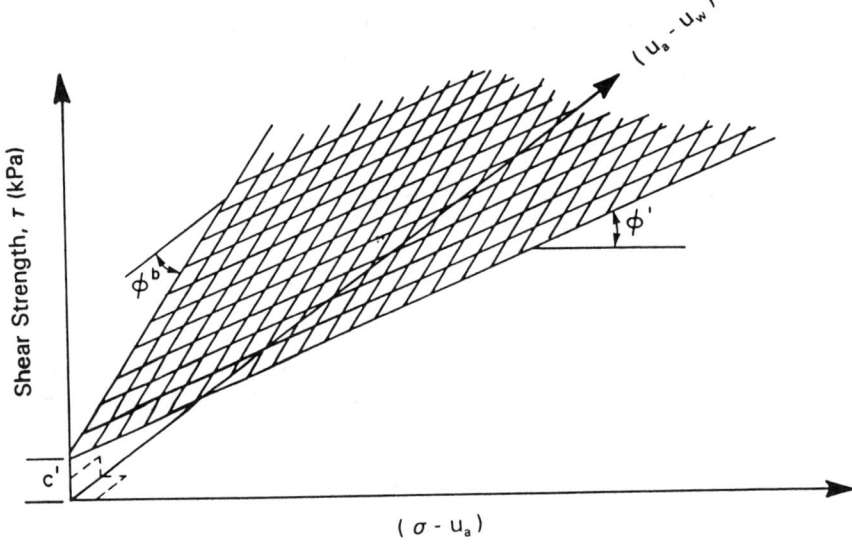

Fig. 13.   A linear shear strength failure surface for an unsaturated soil.

In general, it is possible to linearize the relationship over a selected suction range. At the same time, the Soil Water Characteristic Curve can be used to approximate the shear strength versus suction relationship (Fig. 14). Figure 14 illustrates how the ϕ^b angle begins to deviate from the effective angle of internal friction, ϕ', as the soil desaturates at higher suctions. As the matric suction reaches a value corresponding to the residual water content, the ϕ^b angle appears to approach an angle of zero degrees (or it may even be negative).

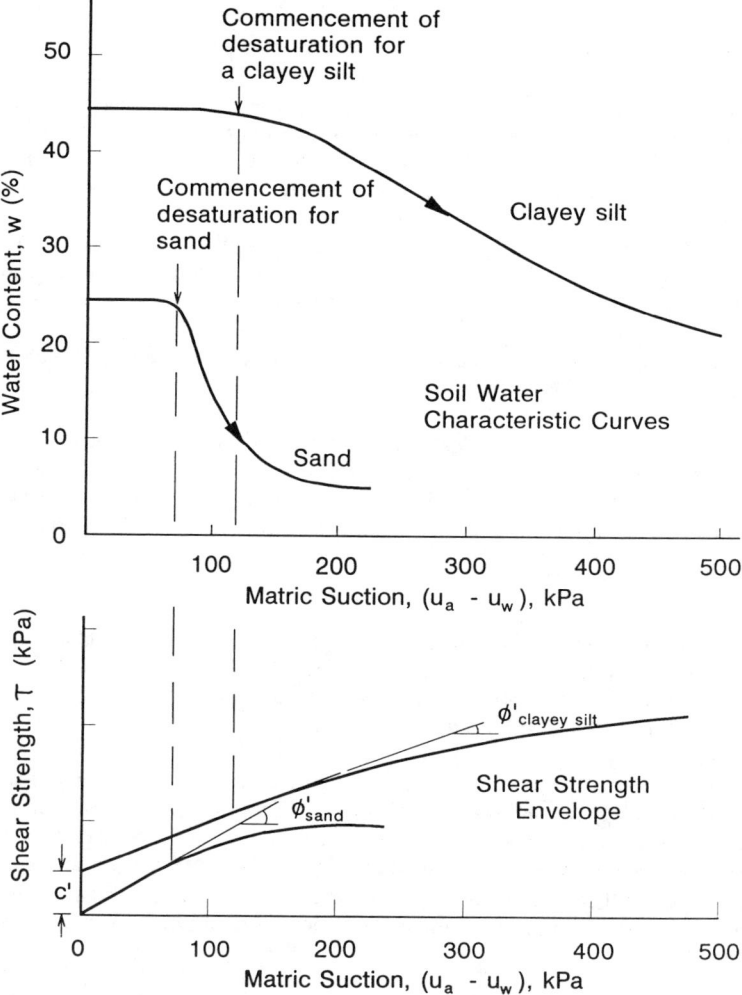

Fig. 14.   Relationship between Soil Water Characteristic curve and shear strength for a sand and a clayey silt

(4)   Coefficient of Permeability:

An unsaturated soil no longer has a constant coefficient of permeability and as a result, the engineer must characterize a permeability function for the soil. Darcy's law still applies in the sense that the velocity of water flow is proportional to the hydraulic head gradient. The water flow equation can be written as follows.

$$v_w = - k_w(u_a - u_w) \, \partial h / \partial y \qquad\qquad [6]$$

The coefficient of permeability is a function of the stress state in the soil. In particular, the coefficient of permeability is a primary function of matric suction. The permeability function can be empirically computed from a knowledge of the saturated coefficient of permeability and the Soil Water Characteristic curve (Fig. 15). The computations are based on the assumption that water can only flow through the water portion of the soil. Therefore, an integration along the Soil Water Characteristic curve provides a measure of the quantity of water in the soil. For most geotechnical problems, this form of permeability function characterization is satisfactory. Several other functions have also been proposed but at present, Gardner's equation (1958) appears to have quite wide acceptance (Fig. 16).

PROBLEMS WHERE UNSATURATED SOIL MECHANICS (AND SURFACE FLUX) PLAY A ROLE

There are many problems routinely encountered in practice where unsaturated soil mechanics and the surface flux boundary condition have a strong influence on the performance of the structure. A few examples are listed below, under the general soil mechanics categories of i.) seepage, ii.) shear strength and iii.) volume change.

SEEPAGE EXAMPLES

1. Compacted clay covers have become a common solution used to control water flow through waste management facilities (Fig. 17). The performance of the cover is largely controlled by permeability and storage characteristics of the cover along with the surface flux to which the cover is subjected.

2. The movement of moisture in the entire region surrounding a waste containment area is often closely related to the coefficient of permeability of the unsaturated soil zone. The mounding of the groundwater table below a waste containment area occurs in response to flow through unsaturated soils (Fig. 17). Although the soil below the waste containment may maintain negative pore-water pressures, there still will be flow in accordance with the unsaturated coefficient of permeability and the hydraulic gradient.

3. There is generally little flow of water during the construction phase, from a compacted earth structure such as a dam. As the reservoir is filled, there will be water flow through the dam both in the negative and positive pore-water pressure

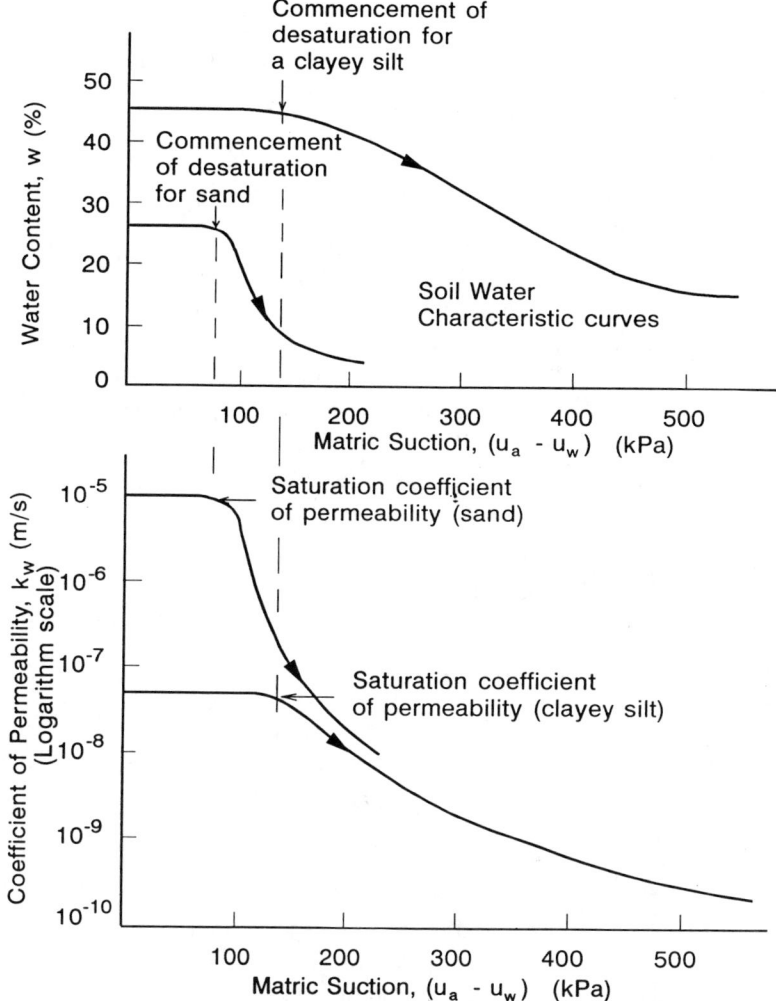

Fig. 15.   Relationship between the Soil Water Characteristic curve and the coefficient of permeability for a sand and a clayey silt

regions. The permeability function must be used when modelling the flow of water through the unsaturated zone.

Once the reservoir is filled, a steady state condition is achieved and most of the flow is through the saturated soil. However, one of the conditions which can trigger instability arises from the extended infiltration on the surface of the dam. In this case,

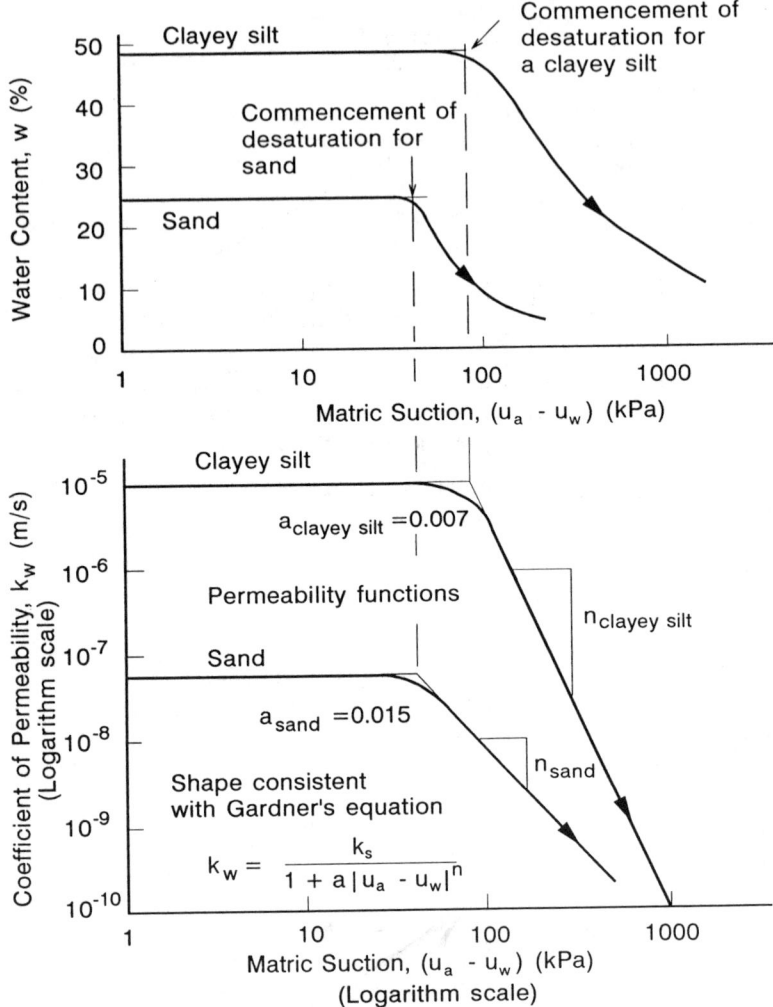

Fig. 16.   Typical Gardner's empirical permeability functions shown for a sand
             and a clayey silt

water will infiltrate through the unsaturated zone above the phreatic line.

4. Expansive soils often cause distress to light structures as a result of the ingress of
water into the soil. The initially high negative pore-water pressures increase and the
soil expands. The flow may be through the fissures and cracks in the soil. This
becomes a difficult unsaturated soil problem to analyze.

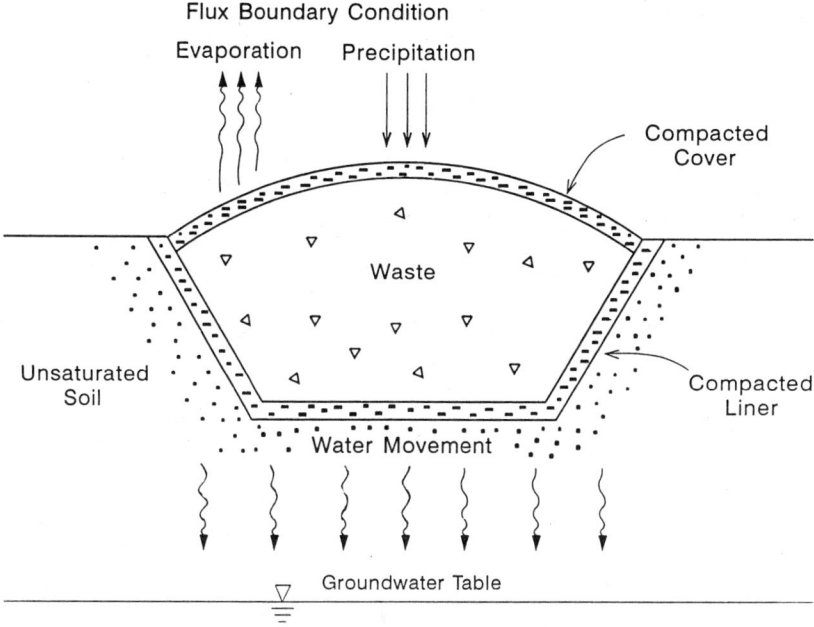

Fig. 17.   An example of the movement of water through a cover as well as flow in the unsaturated zone below a liner.

5. Long-term predictions with respect to wastes resulting from mining operations are strongly controlled by the assessment of the surface flux boundary conditions. This is particularly true for the case of "closure".

SHEAR STRENGTH EXAMPLES

1. Natural slopes generally fail at some time following a high level of precipitation over a prolonged period of time. While the mechanism leading up to failure is well known, few attempts have been made to model the problem. The main reason appears to be related to the difficulties in modelling the flux boundary condition and the flow through the unsaturated soil.

2. The stability of loosely compacted fills can result in high velocity mass movements upon approaching saturation. The soil structure may experience collapse with the result that the load is transferred onto the water phase.

3. The stability of cuts or trenches for laying pipelines involves unsaturated soils and is difficult to assess. The costs associated with temporary bracing are high and each

year lives are lost because of inadequate bracing of the excavations. The stability of temporary excavations is related to the strength of the unsaturated soils.

4. The assessment of the stability of temporary excavations around construction sites is a difficult problem to solve and it is often left up to the contractor to handle in an appropriate manner. It has been observed in Southeast Asia that it is quite common practice for contractors to place a plastic membrane in the vicinity of a newly excavated slope (Fig. 18). The slope may be part of an excavation for a foundation or part of the site remediation (or landscaping). The plastic membrane ensures that a major portion of rainfall will be shed to the bottom of the slope. In other words, the use of the plastic membrane is an attempt to maintain the negative pore-water pressures in the backslope. The practice of using plastic membranes in this manner has potential usage in other parts of the world.

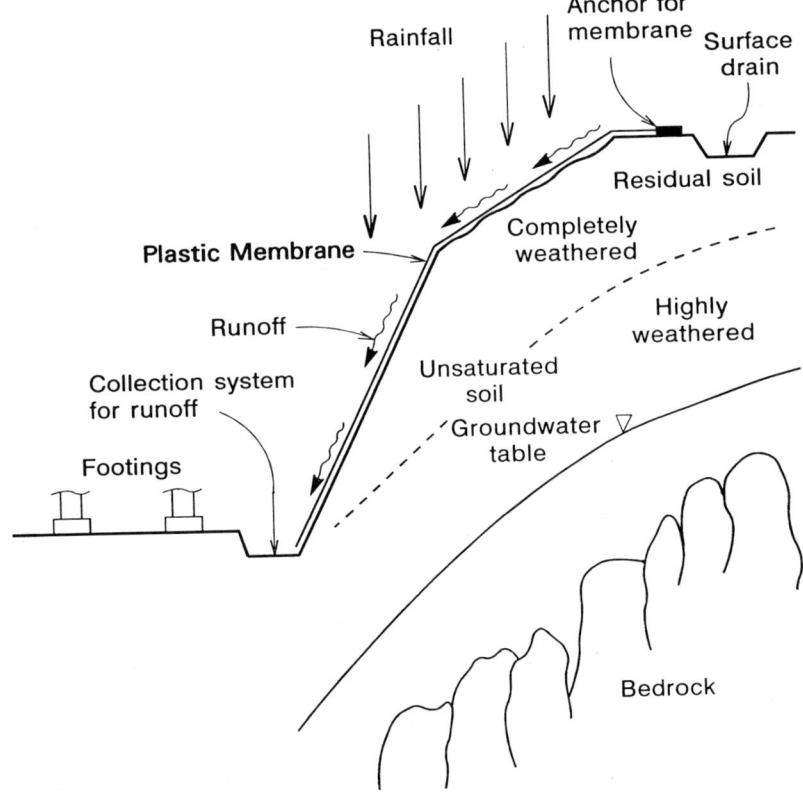

Fig. 18.   Example of the control of infiltration through the use of geomembranes.

5. Engineers realize that it is preferred design practice to use cohesionless material as backfill for an earth retaining structure. However, many retaining structures are backfilled with cohesive materials which change volume in response to the intake of water. The lateral pressures against the wall are a function of the shear strength of the unsaturated soil.

6. The bearing capacity of shallow footings is commonly based on the compressive strength of the soil. These strength measurements are often performed on soil specimens from above the groundwater table where the soil has negative pore-water pressures and may be fissured and unsaturated. The assumption is then made for design purposes that conditions in the future will remain similar. This may not be a realistic assumption.

VOLUME CHANGE EXAMPLES

1. Volume changes in the soil below shallow footings generally take place in response to a moisture flux around the perimeter of the structure. While the mechanism involved is well understood, there is still a need for two- and three-dimensional numerical models to simulate these conditions.

2. In some countries, the shrinkage of high volume change soils under drying conditions poses a more serious problem than swelling. Shrinkage is the reverse of swelling in terms of a change in the stress state. However, the analyses and solutions to the two problems can be quite different.

3. The collapse of meta-stable-structured soils such as loess or poorly compacted silts and sands is a problem involving a volume decrease resulting from a decrease in suction. There is need to have the analysis of collapsing soils embraced within the unsaturated soils theory.

4. The prediction of the depth of cracking is related to the volume change behaviour of the soil. The depth of cracking influences earth pressure and slope stability solutions. An increase in suction increases the depth of cracking. However, a complete analysis of the problem is still not available.

5. There needs to be an analysis for the volume change of compacted fills such as dams and embankments. The changes in volume of the unsaturated soil may be due to total stress or matric suction changes.

6. The design of a soil nailing in retaining structures assumes that load will come on to the anchors as a result of the influx of water into the soil behind the cover on the slope. It is the change in negative pore-water pressures in response to a ground surface flux which initiates the volume change required to load the anchors.

7. While the settlement of trench materials is by no means a recent phenomenon, it

has been a subject of new investigations in recent years. The volume changes (i.e. generally volume decreases) of the backfill material are related to softening in response to the influx of water.

8. The cracking of covers for waste containment areas is an example of a geo-environmental problem for which a volume change analysis is required.

SIMPLIFICATION OF UNSATURATED SOIL MECHANICS FORMULATIONS

Problems involving unsaturated soils often have the appearance of being extremely complex. This problem is, in part, related to the fact that the engineer is not familiar with unsaturated soil analyses. However, the problem is also related to a lack of published simplifications to the formulations. The concepts and theories need to be made as simple as possible in order for engineers to grasp the principles of their application. The formulation of key equations needs to be made simple while still retaining the effect of the primary variables involved.

An examination of the early formulations for saturated soils (and completely dry soils), reveals that early researchers made wise, simplifying assumptions that produced relatively simple formulations. This is readily observed from the classic lateral earth force, bearing capacity, and other formulations. Our need for the simplification of the complex problems was stated by Lytton in his address to the 7th International Conference on Expansive Soils, (Dallas, TX) when he said, "It is a common dichotomy among responsible engineers in practice to demand simplicity in the analysis and design of foundations on expansive soils and at the same time to insist that any design should recognize the complexity in the details required by particular site conditions." He went on to say, "One point to be made from all of this is that no matter how complex the problem, it can be solved once it has been formulated correctly and an adequate computational method is available."
Our knowledge of unsaturated soil behaviour is presently at the stage where basic research has been completed and all fundamental constitutive relations are known. It has been possible to derive relatively complex formulations for various geotechnical problems. As an example, there are available formulations and calculation procedures for slope stability analyses and other problems involving unsaturated soils (Rahardjo and Fredlund, 1991). However, there is still a need to have simplified models and formulations to assist in geotechnical practice.

PRIMARY NEEDS AND DIFFICULTIES

There are still tremendous research needs and the difficulties associated with problems involving unsaturated soils. Some of the key needs are as follows:

(1)   Measurement of negative pore-water pressures (i.e., soil suction): The measurement of soil suction may take the form of being either direct or

indirect. Both types of suction measurements need to be pursued.

(2) More complete information on the Soil Water Characteristic curves: It is of value to accumulate information on the Soil Water Characteristics curves for various soils. This laboratory test should become a standard test on unsaturated soils.

(3) Simplification of the formulations: There are needs for a series of simplified formulations for various problems involving unsaturated soils.

(4) Complete case histories: There is a need for a complete documentation of stresses, soil properties and behaviour associated with seepage, volume change and shear strength problems involving unsaturated soils.

REFERENCES

Aitchison, G. D., (1964), "Engineering Concepts of Moisture Equilibria and Moisture Changes in Soils", Statement of the Review Panel, Ed., published in Moisture Equilibria and Moisture Changes in Soils Beneath Covered Areas, A Symposium-in-Print (Australia), Butterworths, pp. 7-21.

Bishop, A. W., I. Alpan, G. E. Blight, and I. B. Donald, (1960) "Factors Controlling the Shear Strength of Partly Saturated Cohesive Soils", in ASCE Res. Conf. Shear Strength of Cohesive Soils, Univ. of Colorado, Boulder, CO, pp. 503-532.

Casagrande, A., (1936), "Seepage through Dams", J. New England Water Works, vol. 51, no. 2, pp. 295-336.

Fredlund, D. G., (1979), "Second Canadian Geotechnical Colloquium: Appropriate Concepts and Technology for Unsaturated Soils", Can. Geot. J., vol. 16, no. 1, pp. 121-139.

Fredlund, D. G. and N. R. Morgenstern, (1977), "Stress State Variables for Unsaturated Soils", ASCE J. Geotech. Eng. Div. DT5, vol. 103, pp. 447-466.

Fredlund, D. G. and H. Rahardjo, (1993), "Soil Mechanics for Unsaturated Soils", John Wiley & Sons, New York, 560 pp.

Fredlund, D. G. and H. Rahardjo, (1985), "Theoretical Context for Understanding Residual Soil Behaviour", in Proc. 1st Int. Conf. Geomech. in Tropical Lateritic and Saprolitic Soils, (Sao Paulo, Brazil), Feb., pp 295-306.

Fredlund, D. G. and H. Rahardjo, (1987), "Soil Mechanics Principles for Highway Engineering in Arid Regions", Transportation Res. Record 1137, pp. 1-11.

Fredlund, D. G. and H. Rahardjo, (1988), "State of Development in the Measurement of Suction", in Proc. Int. Conf. Eng. Problems on Regional Soils, (Beijing, China), pp. 582-588.

Fredlund, D. G., H. Rahardjo, and J. K. M. Gan (1987), "Non-linearity of Strength Envelope for Unsaturated Soils", Proc. 6th Int. Conf. on Expansive Soils, New Delhi, India, December 1-3, vol. 1, pp. 49-54.

Fung, Y. C. (1965), "Foundations of Solid Mechanics", Englewood Cliffs, NJ: Prentice-Hall, 525 pp.

Gardner, W.R., (1958). "Some Steady State Solutions of the Unsaturated Moisture Flow-Equation with Applications to Evaporation from a Water-Table", Soil Science, Vol. 85, No. 4.

Hamilton, J. J., (1968). "Effects of Natural and Man-made Environments on the Performance of Shallow Foundations". Proceedings of the Twenty-first Annual Canadian Geotechnical Conference, Winnipeg, Manitoba.

International Society of Soil Science, (1963), "News of the Commission, Commission I (Soil Physics)," Soil Physics Terminology, Bulletin No. 23, 7, published by the Soil Science Society of America, Madison, WI.

International Society of Soil Science (1970), "Glossary of Soil Science Terms", published by the Soil Science Society of America, Madison, WI.

International Society of Soil Science, "Glossary of Soil Science Terms", published by the Soil Science Society of America, Madison, WI.

Lytton, R. L. (1992), "Use of Mechanics in Expansive Soils Engineering", keynote address to the 7th International Conference on Expansive Soils, Dallas, Aug. 3-5.

Papagiannakis, A. T. and D. G. Fredlund, (1984), A Steady State Model for Flow in Saturated-Unsaturated Soils", Can. Geot. J. vol. 21, no. 13, pp. 419-430.

Rahardjo, H. and D. G. Fredlund, (1991), "Calculation Procedures for Slope Stability Analyses Involving Negative Pore-water Pressures", in Proc. Int. Conf. Slope Stability Eng. Development, Applications (Isle of Wight, U.K.), pp.

Terzaghi, K. (1936), "The Shear Resistance of Saturated Sands", in Proc. 1st Int. Conf. Soil Mech. Found. Eng. (Cambridge, MA), vol. 1, pp. 54-56.

Wheeler, S.J. and Sivakumar, V. (1992)., "Critical State Concepts for Unsaturated Soil," Proc. 7th Int. Conf. on Expansive Soils, Dallas, Aug. 3-5, vol. 1, pp. 167 - 172

Wilson, G. W. (1990), "Soil Evaporative Fluxes for Geotechnical Engineering Problems", Ph.D. Thesis, University of Saskatchewan, Saskatoon, Canada.

DEFINITION OF SOIL TERMS

σ	total normal stress
σ_c	total isotropic confining pressure
u_a	pore-air pressure
u_w	pore-water pressure
$\sigma - u_a$	net normal stress
$u_a - u_w$	matric suction
τ	shear stress
c	total cohesion (i.e., $c' + (u_a - u_w) \tan \phi^b$)
c'	effective cohesion
ϕ'	angle of internal friction associated with the net normal stress state variable
ϕ^b	angle indicating the rate of change in shear strength relative to changes in matric suction
e	void ratio
G_s	specific gravity
w	gravimetric water content
θ_w	volumetric water content
a_v	coefficient of compressibility
a_t , a_m	coefficients of compressibility with respect to a change in net normal stress and matric suction, respectively
b_t , b_m	coefficients of water content change with respect to a change in net normal stress and matric suction, respectively
v_w	water flow rate
k_s	saturated coefficient of permeability
h_w	hydraulic head of the water phase
k_w	unsaturated coefficient of permeability which is a function of matric suction, (i.e., $k_w(u_a - u_w)$)
ρ_w	water density
g	gravitational acceleration

TRANSMITTED SWELLING PRESSURES ON RETAINING STRUCTURES

Mustafa Aytekin[1], Warren K. Wray, M. ASCE[2], and
C.V.Girija Vallabhan, M. ASCE[3]

Abstract

The main objectives of this paper are to describe a finite element model to simulate the lateral swelling behavior of expansive soils as a function of soil suction change in the soil domain and to compare the results of the numerical model with results obtained by others in a large-scale experimental study. To achieve these objectives, a mathematical model using the well known Finite Element Method, FEM was developed. Also, a computer program called LATEXP2D was developed for the mathematical model and a quadrilateral isoparametric finite element was employed in the analysis. The strain due to swelling is related to soil suction changes within the depth. The large-scale laboratory tests were simulated in the numerical model. The observed and the estimated lateral pressures found by the numerical modeling were then compared to the experimental results. The lateral pressure distributions from the numerical model compared closely with the results of the large-scale experiments. Thus, it was concluded that by developing proper material properties of the expansive soils, the lateral pressures on the retaining structures can be predicted fairly well for engineering purposes.

The Model of Expansive Soils

In general, stress and deformation response of the soil consists of a component due to external loading such as surcharge and foundation loads, and a second component due to the moisture or suction change of the soil mass. The analysis reported herein accounts for both types of loadings. The deformation character of the medium may be classified as expansion (or swelling) with increase in volume upon an increase in the moisture. LATEXP2D considers both stresses due to swelling and stresses due to external loading.

In this study of the state of stress and deformation in an expansive soil, the real soil is replaced by the simplest mathematical model of a finite and linearly elastic medium with a variable with depth value of E_s. Modulus of elasticity is taken as a function of soil's shear strength which varies with the depth of the soil media to simulate the large-scale laboratory tests by Katti et al. (1983) because the distribution of the shear strength of the soil was known.

[1] Assistant Professor of Civil Engineering, Karadeniz Technical University, Trabzon, Turkey
[2] Professor of Civil Engineering, Texas Tech University, Lubbock, TX 79409
[3] Professor of Civil Engineering, Texas Tech University, Lubbock, TX 79409

Strains and Stresses in Expansive Soil

The total strains at each point of an expansive soil subjected to suction variations may be considered as being made up of two components. The first component is a uniform expansion proportional to the suction variations with equal expansion in all directions for an isotropic soil. Thus, there would arise only normal strains and no shearing strains. The strain in any direction for plane strain condition can be obtained from Eq. (2) as follows.

$$\varepsilon_{swell} = \frac{\gamma_h(\Delta pF)}{2} \tag{2}$$

where ε_{swell} = strain due to swelling of soil,

γ_h = suction compression index,

ΔpF = the difference between initial and final soil suctions within the element, in pF units.

The second component of the total strain comprises that required to maintain the continuity of the soil mass as well as that arising due to external loads. These strains are related to the stresses by Hooke's law. The total strains are the sum of the two components. Thus, it is seen that the total strains at each point in an expansive soil consist of two parts: the expansion due to suction change and the strains dependent upon the stress state in the soil mass.

Strain-Displacement Matrix of the Soil-Element

Picture an isoparametric four-node quadrilateral element with the nodes numbered 1, 2, 3, and 4 in counterclockwise sequence. A nonorthogonal intersection of two lines defined as the natural coordinate axes s and t has been superimposed on the element. Interpolation formulas for displacements and global coordinates and the shape functions are the same for each set of equations. The natural s-t coordinates are the independent variables for all equations. The equations (assumed displacement functions) are as follows:

$$\{\varepsilon\} = \begin{Bmatrix} \varepsilon_x + \varepsilon_{swell} \\ \varepsilon_y + \varepsilon_{swell} \\ \gamma_{xy} \end{Bmatrix} = [B] \begin{Bmatrix} u_1 \\ v_1 \\ u_2 \\ v_2 \\ u_3 \\ v_3 \\ u_4 \\ v_4 \end{Bmatrix} + \begin{Bmatrix} \varepsilon_{swell} \\ \varepsilon_{swell} \\ 0 \end{Bmatrix} = [B]\{q\} + \begin{Bmatrix} \varepsilon_{swell} \\ \varepsilon_{swell} \\ 0 \end{Bmatrix} \tag{3}$$

where $\{q\}$ is the nodal displacement vector. The entries due to swelling of the soil in the [B] matrix, which defines the strain in terms of the nodal displacements, will be at the ninth column while the rest of the entries will be due to external loading. The entries in the ninth column of the [B] matrix are

$$B_{19} = \varepsilon_{swell} \qquad B_{29} = \varepsilon_{swell} \qquad B_{39} = 0 . \tag{4}$$

The Soil-Element Stiffness Matrix

Summation of strain and potential energies for a four-node quadrilateral element is given as:

$$\Phi = \frac{1}{2} \int_\Omega \{\varepsilon\}^T [D] \{\varepsilon\} \, d\Omega + \sum_i V_i \qquad (5)$$

where $\{\varepsilon\}$ is known in terms of the nodal displacement vector $\{q\}$, and force potentials are derived for the body forces, swelling forces, and joint loads in terms of the nodal displacement vector and shape functions N_i. The $\sum_i V_i$ term represents total potential energy external loads and their deflections. $[D]$ is the constitutive matrix that is defined for both plane-stress condition, and plane-strain condition (Grandin, 1986).

The stiffness matrix can be written as:

$$[k] = \int_\Omega [B]^T [D] [B] \, d\Omega \qquad (6)$$

The determination of the stiffness matrix, $[k]$, requires integration of the matrix product over the volume, h dA, of the element. Here, two problems must be solved. The first problem is that the matrix $[B]$ has entries that involve ratios of functions of the natural coordinates s and t. The second problem is that the differential area dA can easily be expressed as dXdY, but this integration variable is not the same as the $[B]$ matrix variable, so a change of variable must be undertaken.

In order to solve the two problems, first of all, the integration would be done numerically, and second, the coordinates of the differential area would be changed from the physical X and Y to the natural s and t by application of the Jacobian determinant of the transformation equations for the two coordinate systems. The numerical double-integration and change-of-variable techniques would be used. The result of the derivation is as:

$$d\Omega = h \, dX \, dY = h|J| \, ds \, dt \qquad (7)$$

Therefore, the following equation can be written:

$$[k] = h \int_A [B]^T [D] [B] \, dX \, dY = h \int_{A*} [B]^T [D] [B] \, |J| \, ds \, dt \qquad (8)$$

Equation (8) can be integrated over the area by using Gauss quadrature.

Distributed Body Forces and Equivalent Nodal Loads

In the four-node quadrilateral element, the equivalent nodal forces for the distributed body force and swelling force is defined in symbolic form as:

$$[k] \{q\} = \{Q\}_{NF} + \{Q\}_{BF} + \{Q\}_{SF} \qquad (9)$$

where $\{Q\}_{NF}$ = applied external nodal loads vector,
             $\{Q\}_{BF}$ = force vector resulting from the distributed body force,
             $\{Q\}_{SF}$ = force vector resulting from the swelling of expansive soil,

Body Forces

For the distributed body forces in the four-node quadrilateral finite element, the equivalent nodal force is defined as follows:

$$\{Q\}_{BF} = h \sum_{i=1}^{2} \sum_{j=1}^{2} W_i W_j \left[N(S_i,t_j) \right]^T \left\{ \begin{array}{c} B_x(s_i,t_j) \\ B_y(s_i,t_j) \end{array} \right\} \left| J(s_i,t_j) \right| \qquad (10)$$

The weighting functions, W_1 and W_2, are unity, and the Gauss points are located at s and t equal to ± 0.5773503 for the element used in this study. The distribution of the nodal forces is dependent on the shape of the quadrilateral element.

Swelling Forces

First of all, swelling strains must be determined to calculate the swelling forces of each element. In order to calculate the swelling strains of an element, variations of soil suction in extreme cases (the driest and wettest seasons of an average year) must be known. Once the initial and the final soil suction values are known for every nodal point on an element, average soil suction change for the element can be calculated as follows:

$$(\Delta pF) = \frac{1}{4} \sum_{k=1}^{4} (h_i - h_f)_k \qquad (11)$$

where h_i and h_f are the initial soil suction and the final soil suction in pF, respectively. By using the following equation, swelling forces at nodal points are determined:

$$\{Q\}_{SF} = \int [B]^T [D] \{\varepsilon_{swell}\} |J| \, ds \, dt \qquad (12)$$

where $\{Q\}_{SF}$ = Swelling force vector,
             $|J|$ = Determinant of the Jacobian matrix.

Calculation of Stresses

The stresses in the quadrilateral element,

$$\{\sigma\} = [D][B] \{q\} \qquad (13)$$

are not constant within the element unlike the constant strain triangular element. They are functions of natural coordinates, (s and t), and consequently vary within the element. In here, the stresses are evaluated at the centroid of each element

($s=0$, $t=0$) which are also the points used for numerical evaluation of [k], the element stiffness matrix, where they are found to be accurate.

Simulations of large-scale experiments

The large-scale experiments that were performed by Katti et al., referred to as Katti's experiment hereafter, are simulated in the new finite element model using the computer program LATEXP2D. In the numerical model, modulus of elasticity is estimated by using the known soil properties.

Expansive Soil Backfill Only

In the Katti experiment, black cotton soil, which is thought to be from the Malaprabha Right Bank Canal Km No. 76 (MRBC-76) Karnataka State, India since the properties of the soil match with one another (Katti and Katti, 1987), was used to measure the lateral swelling pressure on the rigid wall which can be assumed to be at an at-rest condition because there would be no lateral movement of the wall. In the experiment, MRBC-76 was placed in the tank in an air dry condition. Then, it was saturated. The lateral pressures were measured using reaction jacks and proving rings placed at 60 cm. (24 in.) vertical intervals.

In the numerical model, modulus of elasticity, E_s, and Poisson's ratio, μ_s, of each element are required to analyze the problem. However, these properties of MRBC-76 were not cited in Katti's work (1983). Therefore, E_s, and μ_s, had to be estimated by using some known properties of MRBC-76. For instance, E_s values were calculated from the design shear strength, c, of MRBC-76. The vane shear strength (c_v) distribution of MRBC-76 with depth, seen in Fig. 1, is used to get the distribution of E_s with depth using an empirical equation given by Skempton and Henkel (1957).

Since the vane shear strength of the soil is given in Fig. 1*, a correction factor, λ, which is a function of plasticity index (PI) of the soil, must be used to get the design shear strength, c, of the soil (Bowles, 1988). Thus, Eq. (14) is used to estimate values of design shear strength. Then, E_s values varying with depth have been calculated using design shear strength, c, values.

$$c = \lambda \, c_v \qquad (14)$$

where c = design shear strength,

               c_v = vane shear strength,

               λ = correction factor, a function of PI ($\lambda = 0.92$ for MRBC-76).

For an expansive clay, a typical value of Poisson's ratio of $\mu_s=0.3$ (Amir and Sokolov, 1980) was used since there was no data from which a value of μ_s could be predicted from Katti's work.

*Katti et al. reported their results in units of pounds per square foot and kilograms (force) per square meter. The authors have retained these units in this paper to permit a direct comparison of predicted and measured results. The reader may convert psf to kPa by multiplying psf values by 96.53 and may convert kg/cm^2 to kPa by multiplying kg/cm^2 by 2.356.

On the other hand, in order to calculate the stresses transmitted to the retaining structure, the computer program LATEXP2D also requires the difference of the initial and the final soil suction distributions over the depth of the soil. In Katti's experiments, the soil was placed in air-dried conditions and then saturated. Thus, it is assumed that both the initial suction and the final suction would be constant over the depth of the soil. The two extreme conditions (air dry and saturated), which are the driest and the wettest states of soil at its surface, have suction values of about 6.0 pF (Russam and Coleman, 1961) and 2.0 pF, respectively (Aitchison and Richards, 1969). Therefore, the maximum difference in change of soil suction value is 4.0 pF (6.0 pF - 2.0 pF). A value of 3.0 pF for the soil suction change was taken over the depth of the soil used in the experimental tests because the soil in Katti's test could not experience the extreme conditions during the tests.

Another parameter which also has to be given to LATEXP2D is the dimensionless coefficient of suction change compressibility, γ_h; the coefficient is a function of the type and amount of clay in the soil. Some authors, e.g., McKeen and Hamberg (1981), refer to this coefficient as "suction compression index", SCI. There are two ways to find suction compression index (γ_h) values. One way is from a set of equations (McKeen, 1977) called strain equations. The other way is from a chart where γ_h is found as a function of activity (Ac), Cation Exchange Activity (CEAc), and Coeffieient of Linear Extensibility (COLE) (McKeen, 1980). The strain equations and the chart give different γ_h values for similar clays. Here, γ_h values are calculated by the strain equations since these γ_h values gave more reasonable results than the γ_h values from the chart. In Katti's experiment, the expansive soil, MRBC-76, had a 55 percent clay content. The value of γ_h for a 55 percent clay soil consisting wholly of smectite is calculated from McKeen's (1977) equations to be 0.026. Since the composition of the clay mineralogy was unknown and since it was unlikely that the soil was pure smectite, the soil was assumed to be a mixture of clay minerals and the value of the coefficient of suction compression was taken to be $\gamma_h = 0.020$.

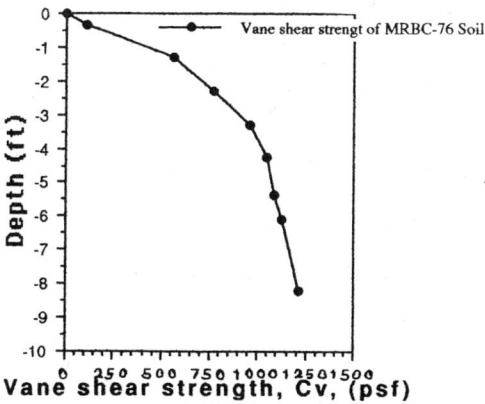

Figure 1. Shear strength distribution of MRBC-76 with depth.

A comparison of lateral swelling pressure of MRBC-76 between the experimental observations and the numerical data is shown in Fig. 2. There is a little difference between the curves from experimental work and the numerical modeling as seen in Fig. 2.

It is believed that one of the reasons for the difference between the predicted and measured pressures is there was likely some unaccounted for lateral deflection in the wall of the tanks that had been used in Katti's experiments. It is well known that even minute displacement of a wall in the lateral direction will result in a very large relief of lateral swelling pressure. However, in the numerical model, the lateral deflections of the tanks have been assumed as zero over the entire depth. In order to see the distribution of lateral pressure for a deflected wall, some lateral displacement (about 2.5 cm or 1.0 in.) at the top of the wall and decreasing with depth was assumed as seen in Fig. 3. Then the lateral pressure distribution was recalculated. The recalculated lateral pressure distribution from the deflected wall in the numerical model and the observed lateral pressure distribution in Katti's test are almost same as seen in Fig. 4.

Another reason for this lateral pressure difference between experimental observations and that of the numerical modeling might be the differences between the dimensions of the tank that was used in Katti's experiments and that assumed in the numerical model. In addition to these reasons, the soil suction change over the depth of Katti's tank was not reported so that the soil suction change may not be a constant in the tank unlike the estimation in the numerical model. Thus, the lateral pressures were found to be larger in the numerical model than in the experimental work.

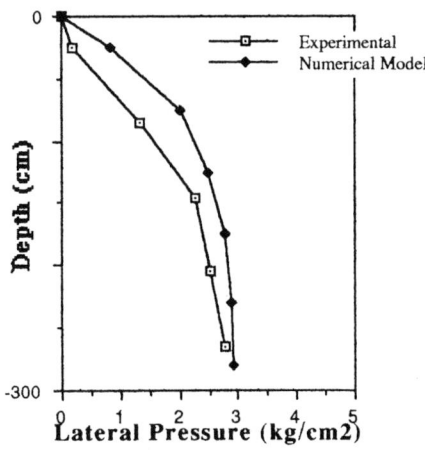

Figure 2. Comparison of transmitted lateral pressures from the numerical model and the experimental investigation for expansive soil only.

CNS Soil Backfill Only

In the cohesive nonswelling (CNS) soil backfill only, Katti used the tank that had been used for the expansive soil only. Thus, in this case of the simulation of CNS soil backfill only in the numerical modeling, the dimensions of experimental setup are taken to be the same as in the case of expansive soil only. In the numerical model, the distribution of the modulus of elasticity of CNS soil (CNS soil of Byahatti) over the depth of the soil is taken as linear with a value of zero at the surface and a value of 8,620 kPa (180,000 psf) at a depth of 10 ft. Unlike the expansive soil backfill only, there was no data from which the distribution of the modulus of elasticity, E_s, with depth could be estimated. A value of Poisson's ratio of $\mu_s=0.3$ is also taken for the CNS soil as it was for the expansive soil.

Figure 5 was plotted by using some relationships between the moisture content and the soil suction. In general, the soil suction is about 6.0 pF in the driest state (Russam and Coleman, 1961). Also, the soil suction values are about 3.3 pF and 0.1 pF for plastic limit and liquid limit, respectively (Croney and Coleman, 1954). Thus, using these soil suction values, and the soil properties of the CNS soil and the expansive soil that are used in Katti's experiments, Fig. 5 was plotted. Using Eq. (15) (Aytekin, 1992), the soil suction is found to be about 4.5 pF for a moisture content of 15 percent as seen in Fig. 5.

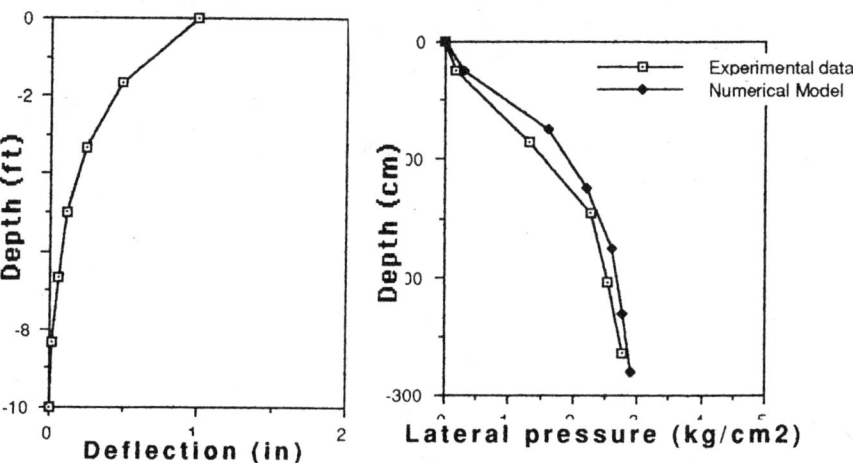

Figure 3. Assumed displacement   Figure 4. Comparison of transmitted
          distribution of the wall in .   lateral pressure distribution
          Katti's experiment   from the experimental
                                                     work and the deflected wall
                                                     in the numerical model.

$$(\Delta pF)_{CNS} = pF_{w_{15}} - 2.5 \, pF \qquad (15)$$

where $(\Delta pF)_{CNS}$ = soil suction change in CNS soil, in pF.

$pF_{w_{15}}$ = soil suction of CNS soil at 15 percent water content, in pF.

Since LATEXP2D requires the soil suction change, which is ($pF_{initial}$ - pF_{final}), over the depth of the soil, there was no need to estimate the initial and the final suction distributions of soil in Katti's experiment to calculate the lateral pressures transmitted on the retaining structures. Instead of assumption of the two soil suction distributions (the initial and the final soil suction) in Katti's experiment, only the estimated difference of these two distributions was required and which was found from Fig. 5 and Eq. (15).

Another parameter, which had to be predicted, was γ_h for the CNS soil. In Katti's experiment, it was reported that the CNS soil had a 35 percent clay component. However, mineralogical composition of the clay was not reported.

Thus, γ_h-values for the CNS soil have been calculated for kaolinite, illite, and montmorillonite as 0.006, 0.013, and 0.015, respectively, using the strain equations (McKeen, 1977).

Therefore, a value for the coefficient of suction change compressibility of 0.006 was taken for the CNS soil because it is expected that the volume change potential of the selected CNS soil for the purpose of Katti's test must have a very low volume change potential with variation of its moisture content.

A comparison of the lateral swelling pressure between the experimental measurements from CNS soil of Byahatti by Katti and the numerical data are shown in Fig. 6. As seen in Fig. 6, the experimental and numerical results are very nearly the same.

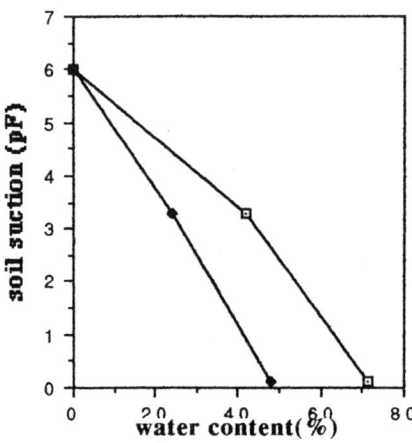

Figure 5. The relationship between the soil suction and the water content for the CNS soil and the expansive soil.

Different Thicknesses of CNS Soil Backfills

Katti tested four different thicknesses of the CNS soil backfills in his large-scale laboratory investigation. The thicknesses of the CNS soil, which Katti called "backing", between the expansive soil and the tank wall were 100 cm, 60 cm, 40 cm, and 20 cm. In this paper, the same thicknesses of backings are simulated in the numerical model. The comparison of transmitted lateral pressures from the numerical model and Katti's experiment for the thickness of 100 cm, can be seen in Fig. 7. As seen in Fig. 7, the experimental and the numerical plots give similar values of transmitted lateral pressure for 100 cm. thickness of CNS soil (or 100 cm. backing).

The values are very close to one another. However, the numerical model gives more conservative lateral pressure values (i.e., larger values) over the depth of the soil with respect to the experimental study for the thicknesses of 60 cm, 40 cm, and 20 cm of CNS soil. The larger difference between the experimental measurements and the results from the numerical model for a maximum lateral pressure occurs for 40 cm backing. The transmitted lateral pressures at a depth of 220 cm (7.22 ft) for 40 cm backing are 1.14 kg/cm^2 and 1.85 kg/cm^2 from the experimental and the numerical model, respectively. The differences between the numerical results and observed results from large-scale experimental study are smaller than this difference for the rest of the cases. The reasons for these differences can be attributed to the differences in the soil suction profile that was used in the numerical model and the actual soil suction profile in the soil of the experiments. Additionally, it is likely that some small lateral wall deflections were present in the experiments permit stress relief in the lateral direction as pointed out above. Also, it is likely that the lateral wall deflections and the soil suction profiles were different in each experiment.

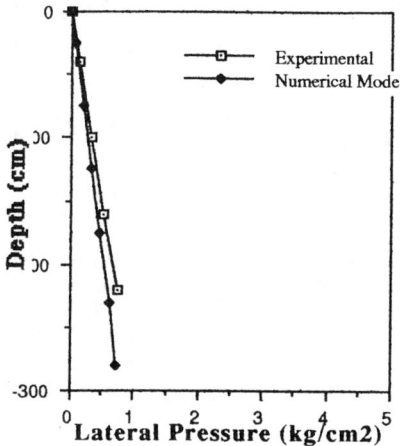

Figure 6. Comparison of transmitted lateral pressures from the numerical model and experimental investigation for CNS soil only.

Conclusions

The results from the numerical modeling and several large-scale laboratory tests performed by Katti et al. compare quite favorably.

In order to estimate the lateral swelling pressure transmitted to a retaining structure by swelling soil, the soil suction envelope must be known over the depth of the active zone. Since the computer program LATEXP2D is sensitive to the modulus of elasticity of soil, determination of the modulus of elasticity distribution in a soil media is important. Also, while determining the moisture content of the soil mass in different locations, these locations must be chosen carefully to represent the soil mass as accurately as possible.

Another very important parameter in the calculation of the transmitted lateral pressure is the suction compressibility index, γ_h, which is a function of the clay content and the mineralogical composition of the soil mass. The clay percentage and the composition of the clay mineralogy must be determined to estimate the lateral pressure as closely as possible. In this study, the γ_h values calculated by the strain equations of McKeen (1977) gave better results than the γ_h values taken from the McKeen chart (1980) with respect to duplicating measured transmitted lateral swelling pressures.

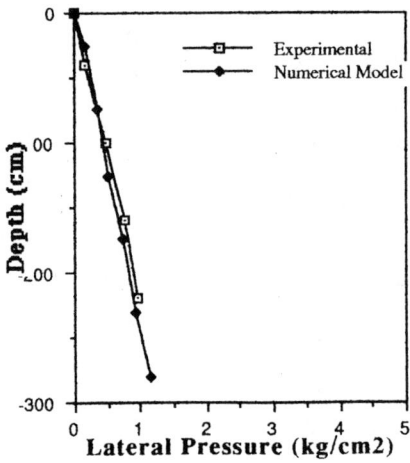

Figure 7. Comparison of transmitted lateral pressures from the numerical model and experimental investigation for 100 cm CNS backfill.

References

Aitchison, G. D., and Richards, B. G., 1969, "The fundamental mechanisms involved in heave and soil moisture movement and the engineering properties of soils which are important in such movement," Proceedings, 2nd international conference on expansive soils.

Amir, J. M., and Sokolov, M.S., 1980, "Analog model for piles in expansive clay," Proceedings, 4th International Conference on expansive soils, Denver, CO, pp. 582-595.

Aytekin, M., 1992, "Finite element modeling of lateral swelling pressure distributions behind earth retaining structures," dissertation presented in partial fulfillment of the requirements for the degree of doctor of philosophy, Texas Tech University, Lubbock, TX.

Bowles, E. J., 1988 "Foundation Analysis and Design", 4th edition, McGraw-Hill, New York, pp. 156.

Croney, D., and Coleman, J. D., 1954, "Soil structure in relation to soil suction (pF)," Journal of Soil Science, Vol. 5, No. 1, pp. 75-84.

Grandin, H. Jr., 1986, Fundamentals of the Finite Element Method, Macmillan Publishing, New York,

Katti, R. D., and Katti, R. K., 1987, "Studies on passive resistance development in saturated expansive soil," Proceedings, 6th international conference on expansive soils, New Delhi, India, Vol. 1, pp. 61-66.

Katti, R. K., Bhangle, E. S., and Moza, K. K.,1983, "Lateral pressure of expansive soil with and without a cohesive non-swelling soil layer-Applications to earth pressures of cross drainage structures of canals and key walls of dams (studies of Ko-condition)," Central Board of Irrigation and Power, Technical Report No. 32, New Delhi, India.

McKeen, R. G., and Hamberg, D. J., 1981, "Characterization of expansive soils," Transportation Research Record No. 790, Transportation Research Board, Washington D. C., pp. 73-78.

McKeen, R. G., 1980, "Field Studies of Airport Pavements on Expansive Clay," Proceedings, 4th International Converence on Expansive Soils, Denver, CO, Vol. 1, pp. 242-261.

McKeen, R. G., 1977, "Characterizing expansive soils for design," Presented at the Joint Meeting of the Texas, New Mexico, and Mexico Sections of the ASCE, Albuquerque, New Mexico, 23 pp.

Russam, K., and Coleman, J. D., 1961, "The effect of climatic factors on subgrade moisture conditions," Geotechnique, Vol. 2, No. 1, pp. 22-28.

Skempton, A. W., and Henkel, D. J., 1957, "Tests on London clay from deep borings at Paddington, Victoria and South Bank," Proceedings, 4th international Conference on Soil Mechanics and Foundation Engineering, London.

NEGATIVE SKIN FRICTION DUE TO
WETTING OF UNSATURATED SOIL

By Richard C. Hepworth,[1] M. ASCE

Abstract

A steel building at a cement plant in Utah was originally supported by spread footings. Shallow wetting of the coarse granular soils supporting the footings resulted in over three inches of settlement. Piles were used to underpin the building. The northeast corner column was supported by two 12 X 54 WF piles driven to depth 98 feet. About two years after the underpinning a deep wetting of the soils at the northeast corner occurred. The result was several inches of additional settlement on this corner of the building and twelve inches of ground subsidence adjacent to it. Excavation around the pile cap revealed that one of the piles was eight inches below the pile cap. A pull out test performed on the failed pile indicated the skin friction force was at least 182 tons. The building corner was underpinned again with three BP 12 X 74 piles driven to depth 145 feet. The upper 50 feet of each new pile was sleeved with an outer casing. Performance has been satisfactory since.

Introduction

A cement plant was constructed near Leamington, Utah, about 100 miles south of Salt Lake City, in 1981-82, Figure 1. Most of the structure foundations were shallow spread footings or mats bearing on a coarse granular deposit. Several wetting incidences occurred during and after construction. Subsequent to wetting, structures were affected by settlement (Hepworth and Langfelder 1988). An overall description of the project was presented in the previous paper. The present paper describes in more detail one interesting aspect of the larger case history. The previous studies showed wetting of the foundation soils was the cause of excessive

---

[1]Geotechnical Engineer, 6859 N. Village Rd., Parker, CO 80134

settlement. Where wetting occurred, settlements amounted to several inches. The magnitude of settlement, while mostly dependent on the degree of wetting, was also influenced by the type of soils and the size of the footings. A large clinker storage silo has undergone as much as 36 inches of settlement.

The then owner, Martin Marietta, developed an active program to underpin critical structures. Kaiser Engineers designed and managed most of the repairs. About 21 million dollars were spent on remedial measures between 1982 and 1989. The plant remained in productive operation throughout practically all of the repair work. The plant is now owned by Ash Grove Cement West, Inc. and is performing satisfactorily.

The finish mill building was underpinned in 1983. In October, 1984, renewed settlement was noted at the northeast corner of the finish mill building. The investigation reported here indicated the probable cause of the renewed settlement and the remedial measures taken to correct the distress.

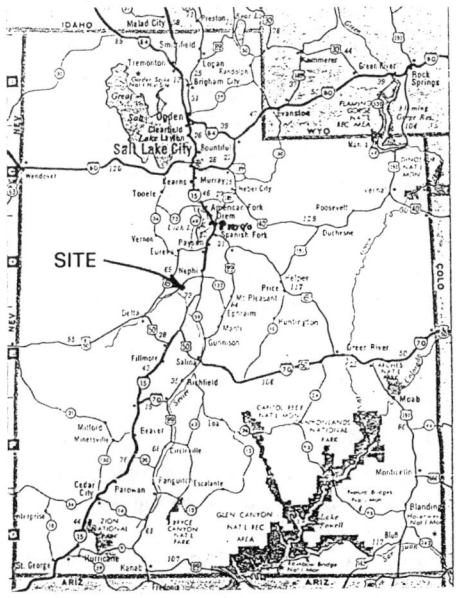

Figure 1. Site Location

Setting

The plant is situated on a northwest facing terrace of ancient Lake Bonneville, south of the Sevier River. The site elevation is 4900 feet which corresponds to the shore line at the Provo stage of Lake Bonneville. A variety of depositional environments existed during and since the lake occupied the area. The resulting sediments consist of lacustrine, near shore sand dunes and coarse colluvial/alluvial deposits. These various sediments were interlayered as the depositional environment changed. The water table is at depth 100 feet, about the level of the Sevier River.

The plant site was prepared for construction by shallow cut and fill to create two benches. The finish mill building is on the lower level at elevation 4885 feet. Surface drainage is provided by interception ditches on the south and culverts and general site grading through the plant. The plant layout is shown on Figure 2.

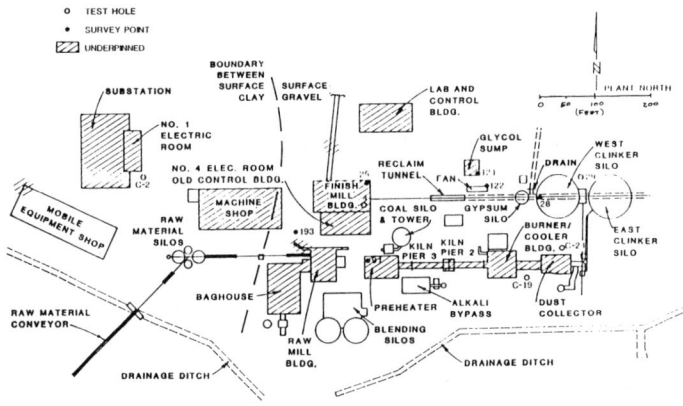

Figure 2. Plant Layout and Mill Building Location

Subsoil Conditions

Many test holes have been drilled throughout the site, by various geotechnical consultants, for design and investigating distress . At the location of the mill building the soils consisted of 35 feet of silty to clayey, sandy gravel; 58 feet of stiff sandy clay; 10 feet of dense silty sand; 37 feet of very stiff sandy clay; and dense silty gravel from depth 140 to 151 feet, the maximum hole depth. A log of the test hole K-28, drilled after failure, is shown on Figure 3. Free water was at depth 101 feet.

Test holes drilled in 1979 and 1980, before construction, showed the soils to have high SPT blow counts, both in the upper granular soils and the underlying clays. Very few moisture tests were performed at that time, but in general, the moisture content was low. Table 1 shows a comparison of tests performed at different times. During the remedial design phase, moisture contents and compression characteristics were specifically studied. On Figure 3 moisture tests are plotted that were taken from borings near the subject area. The plots show moisture of samples taken after initial wetting (1982) and after deep wetting (1984).

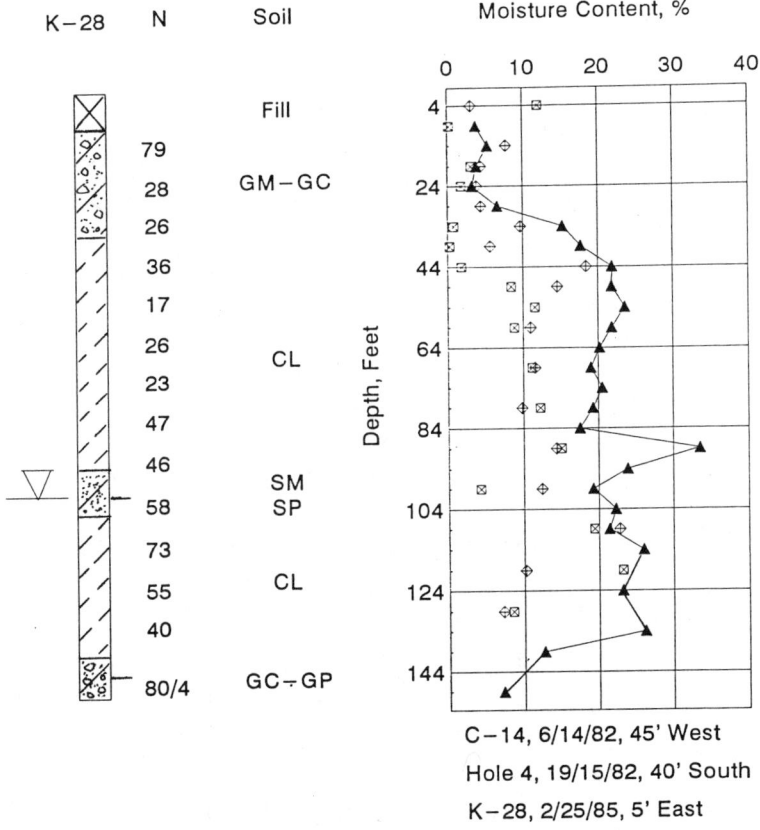

C−14, 6/14/82, 45' West

Hole 4, 19/15/82, 40' South

K−28, 2/25/85, 5' East

Figure 3. Log of Test Hole and Moisture Comparison at Failed Pile

Oedometer tests, Figure 4, showed the fine grained soils exhibit a collapse potential before wetting and showed a high compression index after wetting. Unconsolidated undrained triaxial shear tests, at natural moisture contents, were performed on clay samples taken from depths 50 to 75 feet in Boring K-28, drilled near the failed pile. The shear strength varied from 1.2 to 2.4 ksf with an average of 1.6 ksf. Previous tests performed on the clay showed an average shear strength of about 1.0 ksf.

The upper granular soils represent a rapidly deposited and unsorted material. Lenses of fine sand and inclusions of large gravel, cobbles and occasional small boulders make this deposit very erratic. Obtaining undisturbed samples was difficult because of the large particle sizes. To gain some idea of the compression characteristics of this deposit, samples from test pits were remolded in six inch molds at 3% moisture and 124 pcf dry density. Under a 5000 psf pressure the sample compressed 1.5 % when wetted.

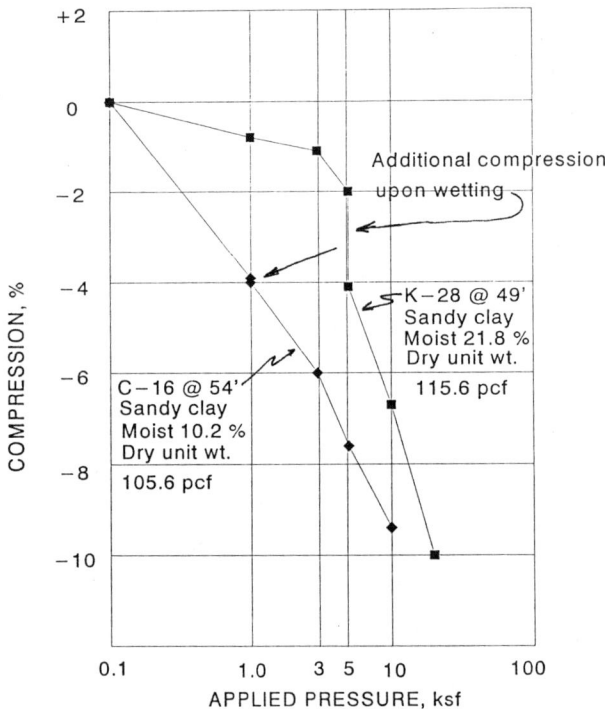

Figure 4. Oedometer Tests Before and After Wetting Incidents

TABLE 1. **Comparative Soil Properties**

Soil Type/Time	N	Range m %	Ave. m %	Dry Dens.	-200 %
Upper Granular					
Before Constr.	30 - >100				
Remedial Study	20 - >100	<1 - 12	6	105 - 120	13-43
Failure Study	20 - 79	3 - 7	5		12-27
Fine Grained Soil					
Before Constr.	28 - 88		14	89 - 106	
Remedial Study	28 - 61	8 - 15	11	104 - 112	60-90
Failure Study	13 - 57	17 - 33	20	100 - 110	62-94

Note: Before Construction; before 1982.
Remedial Study; 1982 to 1984.
Failure Study; November 1984 to June 1985.

The results of the initial remedial design studies showed the upper dense granular soil would compress a small amount upon wetting and the lower fine grained soil had a collapse potential. Settlement readings and observations indicated that if a general wetting occurred, such as from poor surface drainage, settlements would generally be small, but continue for a long time. In contrast, the result of a point wetting source, such as a water line break, settlement would be large and sudden. Where wetting was limited to the upper granular soils, less than six inches of settlement occurred. Where wetting extended to the lower fine grained soils, much larger settlements occurred.

Remedial Design

Over the course of several years most of the structures were underpinned. However, not all structures were affected by excessive settlement and these were not underpinned. Underpinning was undertaken on some structures and machinery because of their low tolerance for differential movement. Because settlement had occurred at some locations at the plant, monitoring points were established on all structures soon after construction.

The mill building was a steel frame and metal clad structure founded on spread footings on the upper granular soils with an allowable bearing pressure of 7000 psf. The total design load for the column at the northeast corner was about 240 kips. Settlement began about one year after the building was completed. As underpinning was in progress at other plant locations, it was decided to provide new support for the mill building. Approximately 3.5 inches of settlement occurred before the northeast corner column was underpinned. Figure 6 shows the settlement history of this corner.

The corner was underpinned in March, 1983, with two 12 X 53 WF piles driven to 98 feet. Driving criteria was to an ultimate bearing capacity of 200 tons for each pile. The ultimate capacity consisted of a working load of 60 tons, a safety factor of two, and 100 tons allowed for negative skin friction. The additional 20 tons needed to meet this criteria would be developed in "freeze up". Negative skin friction was based on an adhesion value of 1.0 ksf for the granular soil, 0.4 ksf for the clay, and a wetting depth of 70 feet. Driving criteria and bearing capacity were established by a WEAP analysis and by using the Case-Goble dynamic analyzer at nearby piles. The latter indicated 90% of the resistance developed below 80 feet. The piles were spliced at 40 foot lengths with a butt weld. A Conmaco 115 hammer was used. Based on a driving energy of 37.4 kip-ft the WEAP analysis indicated bearing capacity would be obtained at 110 blows per foot. Based on the dynamic analysis the criteria were modified to 100 blows per foot or 80 blows per foot for two consecutive feet. During driving the blow count increased rapidly for the last three feet before reaching the criteria.

After underpinning, the area was paved with 12 inches of concrete. Settlement at this corner was about one inch over the next 18 months. In November, 1984, the settlement readings showed a sudden increase and the concrete crash wall and adjacent pavement showed increasing distress. Four inches of additional settlement occurred before jacking was done to stabilize this column.

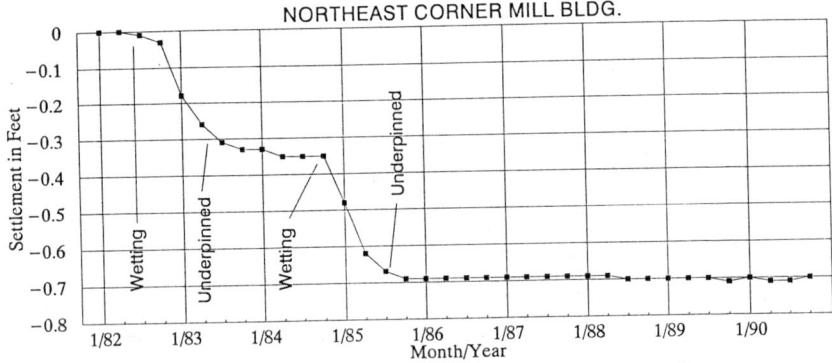

Figure 5. Settlement History, Northeast Corner Mill Building

Failure Study and Reunderpinning

Such rapid movement was a concern because design and pile driving criteria for most previous underpinning had been constructed using similar criteria. A study was undertaken in to determine the cause. Several test pits were opened up at this corner and in adjacent areas and a deep test hole was drilled near the east pile at this corner. Observations made in the test pits confirmed suspicions that a significant point wetting had occurred. It was learned that an accidental overflow of water occurred during October - November, 1984, in the room adjacent to this corner when an automatic shutoff failed.

A void of six inches was found between the pavement and the subgrade in mid-November, 1984, shortly after the renewed settlement was noted. Two months later when large test pits were opened the void had increased to 12 inches. The joint between the pile cap and the pavement was tight and the cap was supporting a large pavement area. The soil around the piles was very wet. Moisture was visible as a coating on the larger soil particles and the fine grained matrix was wet. Moisture content determinations are less indicative of wetting than visual observation of these coarse unsaturated soils. When the bottom of the pile cap was exposed, the top of the east pile was observed to be about eight inches below the base of the cap. This amount of separation indicated that the pile had moved down at least 12 inches.

The moisture determinations from samples taken from Boring K-28 are plotted on Figure 3, and show that wetting around the pile extended through the full thickness of the fine-grained stratum. A dense sand layer from 93 feet to 103 feet was encountered in the test hole. The pile tip was embedded in this stratum. The water table was measured at depth 101 feet. Using the SPT method and a California sampler, blow counts were 46 at 94 feet and 43 at 99 feet in this layer. Based on nearby holes drilled during the previous remedial studies, this was the expected bearing stratum.

Although the evidence of a failed pile was clear, it was not all that believable. It was concluded that wetting caused compression of the fine-grained layer which, in turn, caused the weight of all the soil and adjacent concrete pavement above to be carried by the piles. This increased load exceeded the bearing capacity of the one pile. Also considered was the possibility that the pile had parted at one of the splices. A pull out test was programmed to help in the analysis. The reaction for the load test was to be new piles planned for reunderpinning the building corner.

The new underpinning was designed using three HP 12 X 74 piles. The upper 50 feet of each pile was cased with a 20 inch diameter pipe inserted in a predrilled hole. The void between the pile and the casing was backfilled with a bentonite slurry. The piles were driven to an ultimate capacity of 235 tons in the dense gravel below depth 140 feet. A WEAP analysis indicated 120 to 140 blows per foot for a Del Mag 3002 hammer at 65% efficiency would be required. It was estimated that sleeving to depth 50 feet would reduce the negative skin friction to 50 tons and give a calculated factor of safety of 2.6 for downdrag and structural load. Negative skin friction was calculated using a value of 0.7 ksf adhesion in the clay. This value was based on an average shear strength from the triaxial tests reported above, and considering that much of the collapse had occurred. To eliminate any additional load contribution, the remaining former pile was cut free from the cap.

Load Test

The new piles were driven to the desired criteria and a setup was made to load test the failed pile. The test used two new piles for the reaction. A hole was cored through the old cap and an extension welded onto the failed pile. The extension was welded to a beam spanning to the reaction piles and two 100 ton jacks between the new piles and the beam applied the tensile load. Dial gauges were set on the reaction piles and on the test pile. Loading was done in 50 ton increments. Each increment was maintained until the dial gauge reading was smaller than 0.001 inch per minute.

Near the maximum jack capacity, 182 tons, the load was held for 32 minutes when the weld at the extension broke. Total displacement was 0.4 inch. The load displacement plot is shown on Figure 6.

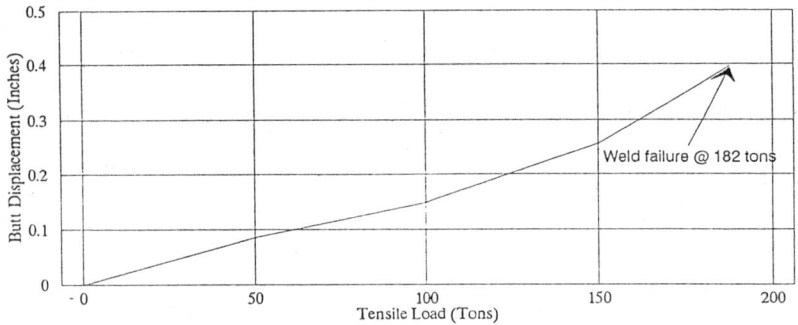

Figure 6. Pile Load Test Results

Analysis

The pullout test indicated total skin friction resistance along the pile length exceeded 182 tons, neglecting the 2.5 ton weight of the pile. Assuming that at least 200 tons would have been required to fail the pile, the average unit skin friction along the 95 feet of effective embedment would be 1.1 ksf. Under this assumption, the total negative skin friction load was two times and the unit skin friction was almost three times the values used for the original remedial design. Wetting was 30 feet deeper than originally assumed. This value also was higher than had been used in the calculation of the redesign. The latter was considered satisfactory as the safety factor for negative skin friction, separate from the structural load, was about three.

Bowles (1988) describes several methods to evaluate skin friction. Back calculating using the α method and the average shearing strength reported above of 1.6 ksf, gives a value for α of 0.7. This value is on the high end of typical values shown by Bowles (1988) and higher than the reduction factor used originally. It should be recognized that in situ shear values and adhesion are best estimates.

Conclusion

The pile failure and load test result showed that negative skin friction is a significant force and can be easily underestimated. The other piles placed under the original remedial criteria are performing satisfactorily. In two other areas pile supported structures have experienced as much as two inches of settlement, but with no severe distress. These larger than expected settlements also are attributed to negative skin friction. The remaining pile at the subject site, placed with the same criteria and soil conditions, did not fail. Prudent judgement and careful weighing of the owner's risk tolerance is needed in remedial design. In the case presented here, failure could have been prevented by assuming full depth wetting, high skin friction values and higher capacity piles with sleeves. Of course these criteria and assumptions would have to be extended to the entire underpinning project and do not appear, in retrospect, to have been warranted. During the course of the remedial work, a large amount of unique data were collected and an appreciation of some uncommon conditions was gained.

Appendix I. References

Bowles, J. E. (1988) Foundation Analysis and Design. McGraw-Hill Publishing Company, New York, N.Y.

Hepworth, R. C., and Langfelder, J (1988). "Settlement and Repairs to Cement Plant in Central Utah" Proceedings: Second Intl. Conf. on Case Histories in Geotechnical Engineering, June 1-5, 1988, St. Louis, MO.

A Laboratory Procedure for Partial-Wetting
Collapse Determination

W.N. Houston[1], H.H. Mahmoud[2], and S.L. Houston[3]

Abstract

Partial wetting results in partial collapse, which is linked to the reduction
in soil suction caused by wetting. The laboratory procedure presented in this paper
represents a modification to the one-dimensional response-to-wetting test. These
modifications are designed to provide quantitative data that can be used to estimate
collapse settlements resulting from partial wetting. The sequence of wetting and
loading, and its effect on the partial collapse data, are discussed. Laboratory data
are presented. Application to engineering practice is addressed.

Introduction

Collapsible soils are commonly identified by performing laboratory
response-to-wetting tests on undisturbed specimens. The conventional laboratory
test procedure is to load the specimen to the stress level expected in-situ, give the
soil free access to water, and then observe the settlement (or swell). The
specimen, when inundated, will experience its maximum collapse potential for the
applied stress. Although maximum collapse potential is of interest, actual field
collapse will often be less than this value because the degree of saturation attained
in-situ may be less than that attained in a conventional laboratory test.

[1]Professor of Civil Engineering, Arizona State University, Tempe, AZ 85286-5306

[2]Senior Staff Engineer, Woodward-Clyde Consultants, Santa Ana, CA 92705

[3]Associate Professor of Civil Engineering, Arizona State University, Tempe, AZ
85287-5306

Partial collapse has been linked by numerous researchers to the reduction in soil suction caused by wetting. Tadepalli, et al. (1992) measured matric suction directly on laboratory specimens while wetting and collapse were occurring. The soil suction values exhibited were very low making, it possible to use a tensiometer for the measurements. They were able to fit a model to the experimental data and describe the collapse process in terms of their model parameters. Other researchers (Alonso, et al., 1990; Kohgo, et al., 1992; and Wheeler and Sivakumar, 1992) have proposed elastoplastic models for the collapse process. Typically, these models include, as a subset, the progressive collapse which occurs as a function of reduction in suction due to wetting.

The authors (Houston, 1992; and Houston, et al., 1993) have addressed partial collapse due to partial wetting in connection with analysis of settlements of foundations. The authors' work to date on partial collapse has been devoted primarily to experimental quantification and application of the results as opposed to model development.

The subject of this paper is a proposed procedure for measuring partial collapse in the laboratory. Before the presentation of the testing procedure is made, a discussion of the mechanism of soil collapse will be given. This discussion is needed to provide a context for the test procedure and a basis for justifying some of the steps in the procedure.

Mechanisms of Soil Collapse

Collapsible soils have been traditionally described as loose, primarily granular soils with varying amounts of dry clay, silt, or other cementing materials holding the granular particles together. A sketch showing the probable fabric of such a soil may have been first introduced by Casagrande (1932) (See Figure 1). These definitions and descriptions of collapsible soils have been repeated and amplified by Dudley (1970) and other researchers, including the authors.

Embankments of compacted clayey soils have been historically known to experience settlements (collapse) upon wetting but have not traditionally been thought of as collapsible. However, in the last one or two decades, several high embankments of this type, particularly in Southern California, have experienced collapse and gained considerable attention (Alwail, et al., 1992). These events have prompted many to take a broader view of the definition of a collapsible soil, attempting to develop some type of unifying theory or model to encompass soil collapsibility for all soil types and conditions.

An attempt to provide a unifying sketch describing the response of various soil types to wetting is given in Figure 2. Figure 2(a) shows qualitatively the

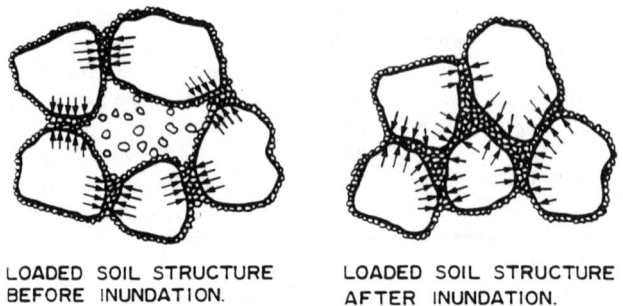

LOADED SOIL STRUCTURE
BEFORE INUNDATION.

LOADED SOIL STRUCTURE
AFTER INUNDATION.

Figure 1. Silt/Clay Structure Suggested by Cassagrande (1932) Before and After Inundation

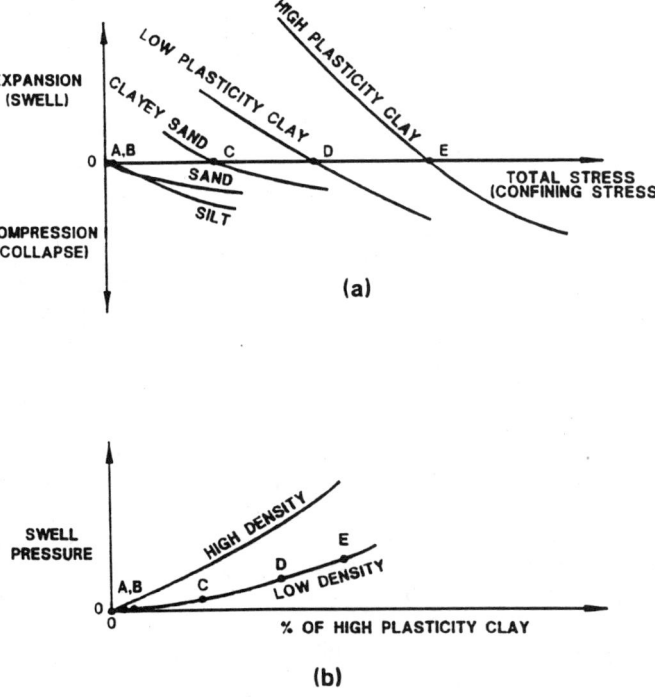

Figure 2. Qualitative Responses of Soils to Wetting

curves relating the amount of swell or collapse upon wetting to the confining pressure (total stress).

The curves for sand and silt in Figure 2(a) are shown as crossing because silts are generally slightly more collapsible than sands, but typically exhibit a slightly higher swell pressure. There are many exceptions to this generality. Regardless of these exceptions, sands and silts both exhibit a very low swell pressure and a very low free swell (under zero confining pressures). Consider, for example, the case of a loose sand with 3 percent dry clay coating the granular particles and acting as a cementing agent. When this soil is wetted, the dry clay coatings will swell as the suction is reduced; but the swelling (and softening) of the clay coatings will allow the granular particles to slide into a denser packing under the influence of any significant confinement, thus resulting in an overall net densification of the mass. The quantity of clay (3 percent) is simply too small for its swelling response to control the overall behavior. Even under zero confinement, the overall swelling of the sand with 3 percent clay is likely to be too small to conveniently measure. If loose sand or silt is cemented with calcium carbonate, for example, wetting will likewise produce nearly immeasurable swell at zero confinement and some collapse at any significant confinement.

Considering the high plasticity clay curve in Figure 2(a), point E represents the confining pressure which is just sufficient to prevent expansion upon wetting, or the zero-strain swell pressure. Obviously, the swell pressure for a low plasticity clay, Point D, would be less. Figure 2(b) shows the cross-plot at zero-strain of the Figure 2(a) data. The swell pressure, given by Points A, B, C, D, and E, increases with percent of high plasticity clay. Figure 2(a) shows that almost any soil can be either expansive or collapsible, depending on the confining pressure at the time of wetting. Even a highly plastic clay can be collapsible if the confining pressure at the time of wetting is high enough.

When one of the clayey soils is wetted with confinement in excess of its swelling pressure, collapse occurs. The mechanism is somewhat similar to that of the case of the clay-coated loose sand discussed above. The relatively dry clay clods behave, to a degree, like granular particles. Upon initial wetting, the outer surfaces of these clods become softened and will allow the clods to slide into a denser packing under the confining stress. When the clay content is high and the confinement is below the swelling pressure, then wetting produces significant swelling.

The sketches in Figures 2(a) and 2(b) are simplified and only qualitative. The positions of the curves vary with soil type and gradation. Also, the curve for a particular soil does not have a fixed, unique position on the plot. Its position depends on the soil density. This dependence is shown qualitatively for the swell pressure in Figure 2(b). In addition, the position of the curve depends on initial water content and maximum past total stress to which the soil has been subjected.

This dependence upon the maximum past total stress has been portrayed as analogous to an expandable yield surface for plastic materials (Alonso, et al. 1990).

Figure 2 can now be used to revisit the traditional definitions of collapsible and expansive soils. It is common for practitioners to define collapsible soils as those which are collapsible upon wetting at ordinary confinement pressures imposed by structures such as overburden and buildings or fills. In addition, practitioners usually think of collapsible soils as those which exhibit nearly immeasurable swell pressure or free swell. More recently, however, there has been some movement toward inclusion of some of the more clayey soils in the collapsible category, provided they exhibit collapse under ordinary foundation and overburden pressures. Likewise, swelling soils have been defined traditionally as those which swell under ordinary foundation pressures. In addition, swelling clays are usually thought of as having significant plasticity.

These traditional definitions would probably be satisfactory if we could agree on what ordinary pressures are; and if these pressures plotted in a fairly narrow band on Figure 2, we could then compare the band with the curve for the soil in question and classify it as collapsible or swelling (or perhaps neither). Unfortunately, however, the band is not narrow and may change with time. The problems with the deep clay fills in Southern California, cited earlier, represent a case where the confinement pressure was significantly higher than values in most engineers' experience.

There are many cases in which there are legitimate concerns over what a soil's response to wetting will be, particularly in the absence of hard data from response-to-wetting tests. In these cases, it is advisable to perform response-to-wetting tests to quantify this response. The tests may be either laboratory tests or in-situ field tests. Typically, the costs of such tests are comparable to the costs of index testing for classification. In any case, the costs are likely to be much less than the costs of misjudging the response to wetting.

Proposed Laboratory Testing Procedure for Partial Collapse

Because the degree of saturation during field wetting of a prototype rarely, if ever, reaches the 95 to 99 percent, which is typical for fully inundated lab test specimens, it is desirable to quantify the partial collapse which occurs in response to partial wetting. As a first attempt to develop a procedure, the authors simply added limited measured amounts of water to a conventional collapse test specimen already under load. It was found that the soil adjacent to the wetting surfaces (the porous stones) was overly compressed and that the densification profile was non-uniform. This result was no doubt due to the temporary reduction in soil suction near the wetting surfaces during wetting. For example, if sufficient water was added to bring the average degree of saturation from 5 percent to 20 percent, the degree of saturation near the stones might temporarily be 80 or 85 percent. If the total stress anticipated for the field was acting on the specimen, the zones wetted to 80 or 85 percent saturation would collapse to a degree commensurate with 80 or 85 percent saturation, rather than the eventual 20 percent saturation.

An alternative testing procedure, which was finally adopted, involves removing the applied load during the addition of each increment of water and allowing equilibration of the degree of saturation before the load is reapplied. The objective of this step is to reduce the load during wetting to a value near the swell pressure shown in Figure 2 so that volume change during partial wetting and equilibration is near zero. Also, the swell pressure for the granular soils being tested was near zero; therefore, it was appropriate to reduce the load down to only that imposed by the porous stone on the specimen.

The test procedure adopted is equally appropriate for undisturbed samples of natural soils or for compacted specimens simulating a compacted fill. The step-by-step procedure commences at the point where the specimen is already in the oedometer ring at natural water content and density, ready for testing:

1.   A sample is prepared for testing in the same manner as in a conventional response to wetting test.

2.   Initial data, such as the weight and diameter of the ring, weight of the soil in the ring, weight of top and bottom porous stones, etc., is recorded.

3.   Filter papers having the same diameter as the porous stones are placed on both the top and bottom of the specimen. The specimen is then set on the bottom stone.

4.   The top stone is then set on the specimen, and a seating load is applied.

5.   The dial gage assembly or LVDT is put into position and zeroed. Note: The dial gage should be attached to an assembly which allows for the removal and placement of the specimen without affecting the reference readings.

6.   A confining stress comparable to that expected in the field is applied to the specimen, and the dry strain (due to dry compression) is recorded in the same manner as in a typical response to wetting test.

7.   The loads are then removed along with the ring assembly (i.e., ring containing the specimen, porous stones, and filter paper). The ring

assembly is then placed on a clean glass plate of known weight and the total weight is determined.

8. A known quantity of water is introduced at various points on the top stone. The ring assembly is then covered with a plastic wrap, to impede evaporation, for a period of approximately four hours to allow the water to be distributed uniformly throughout the specimen (i.e., to achieve a uniform degree of saturation). The time required for the water to be distributed uniformly throughout the specimen may vary for different soil types. The four hours suggested here were adequate for the silty collapsible soils tested.

9. The plastic wrap is removed and the weight of the ring assembly is recorded. The top stone is weighed separately to determine the amount of water it absorbed and the amount of water absorbed by the soil and bottom stone. It is typical that the amount of water absorbed by the bottom stone is negligible.

10. The top stone is put back in the same position, and the ring assembly put back into the loading apparatus at the same orientation. The stress applied in Step 6 is re-applied to the specimen and the collapse strain is recorded as before.

11. Steps 7 through 10 are repeated until the specimen is nearly saturated (i.e., until no additional collapse is observed for an increase in water content).

12. The specimen is then placed in the oven and its dry weight is determined. Using this dry weight, the void ratio and water contents are calculated for each increment of wetting. The degrees of saturation are then calculated using a known value of specific gravity.

Note: The collapse strain at each stage is calculated by subtracting the dry collapse strain from the total collapse strain.

The percent of total collapse is calculated by dividing the collapse strain, corresponding to a certain degree of saturation, by the full collapse strain (obtained when the sample is fully inundated).

Typical Results

The proposed testing procedure was performed on three samples of collapsible silt from Scottsdale, Arizona. All three soil samples were classified as a sandy silt (ML) with Plasticity Index (PI) values ranging from 12 to 15. The results are shown in Figure 3 as dashed curves. Although the suction was not measured while collapse was in progress, it was later determined separately as a function of degree of saturation. The variation of matric suction with degree of saturation is shown as a solid curve in Figure 3 and is considered to be most appropriate for samples 1 and 2. The link between collapse and loss of suction is quite clear, as has been the case for data presented by other researchers.

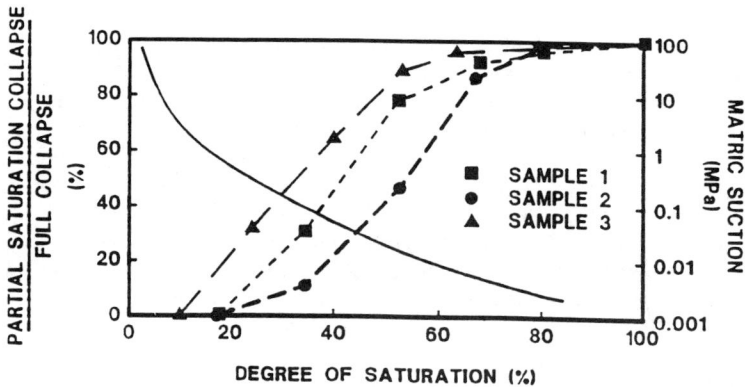

Figure 3. Partial Collapse due to Partial Wetting Curves for Three ML Soils

For a given soil, the position of the partial collapse curve is dependent upon the initial water content, density, confining stress, and maximum past total confining stress applied. In comparing one soil to another, the position of the curve depends on the soil gradation, plasticity, and degree of cementation. It thus appears that if a rather precise estimate of partial collapse is needed, testing of the individual soil may be warranted. If a rough approximation is adequate, it may be satisfactory to select a partial collapse curve corresponding to the gradation and index properties of the soil in question. This option may become more appealing when a larger database of partial collapse responses is published.

Application of Results

If the primary objective is to predict the amount of collapse strain which will occur in the field due to anticipated wetting to an approximate degree of saturation, then application of the results is straight forward. The anticipated degree of saturation is used to enter the measured or adopted partial collapse curve

(Figure 3), and the partial collapse percentage is used to adjust the value of full collapse determined from conventional tests employing full inundation.

If the objective is to obtain parameters for a collapsible soil model or to develop such a model, then it may be appropriate to measure soil suction together with partial collapse and degree of saturation. Depending on the characteristics of the model, the collapse strain may then be quantitatively related to soil suction and other variables. Application to the field might involve use of soil suction, degree of saturation, or other variables as independent variables, with collapse strain being the dependent variable.

As an alternative to the laboratory testing approach prescribed in this paper, the authors have recently developed a down-hole in-situ collapse test system. It is inappropriate to discuss this system in detail here because its features and use will be published at a later time. However, it is relevant to note that the field wetting procedure employed for the test has consistently resulted in degrees of saturation in the wetted zone that are quite comparable to the degrees of saturation resulting from field ponding around model footings of moderate size. It may be that if this in-situ test, or some other in-situ test, can be employed in the field with degrees of wetting imposed that are comparable to the prototype, then no correction for reduced degrees of wetting (as described in this paper) would be necessary.

Conclusions

1.   Partial wetting results in partial collapse, which is closely linked to the reduction in suction caused by wetting.

2.   The position of the partial wetting curve depends on numerous factors including initial density and water content, soil gradation and plasticity, degree of cementation, applied total stress, and maximum past applied total stress.

3.   The general shape and form of these partial wetting curves are expected to be similar to those shown in Figure 3.

4.   The time and cost of performing a partial wetting test is of the order of three or four times that of a conventional laboratory collapse test. Nevertheless, it is feasible and can be performed with ordinary apparatus, when the need for the data is significant.

5.   When the objective is to use partial wetting data to refine an estimate of soil collapse in the prototype, the application of the partial collapse test results is simple and straight forward.

Acknowledgements

The authors wish to thank Mostafa El-Ehwany and Ann Marie Wagner who performed the soil suction measurements and some of the partial wetting collapse tests. This material is based on work supported by the National Science Foundation under Grant No. BCS-89-00 838.

Appendix. References

Alwail, T., Ho, C., and Gragaszy, R. (1992). "Collapse mechanisms of low Cohesion Compacted Soils," *Bull. of the Assoc. of Engr. Geol.*, XXIX(4), 345-353.

Alonso, E.E., Gens, A., and Josa, A. (1990). "A constitutive model for partially saturated soils," *Geotechnique*, 40(3), 405-430.

Casagrande, A. (1932). "The structure of clay and its importance in foundation engineering," *J. Boston Soc. of Civil Engrs.*, 19(4), 168-209.

Dudley, J. (1970). "Review of collapsing soils," *J. Soil Mech. and Foundation Div.*, ASCE, 96(3), 925-947.

Houston, S. (1992). "Partial wetting collapse predictions," *Proc. of the 7th Intl. Conf. on Expansive Clays*, Dallas, TX, August 3-5, Vol. 1, 302-306.

Kohgo, Y., Nakano, M., and Miyazaki, T. (1992). "Soil collapse prediction using an elastoplastic model," *Proc. of the 7th Intl. Conf. on Expansive Clays*, Dallas, TX, August 3-5, Vol. 1, 55-60.

Tadepalli, R., Fredlund, D., and Rahardjo, H. 91992). "Soil collapse and matric suction change," *Proc. of the 7th Intl. Conf. on Expansive Clays*, Dallas, TX, August 3-5, Vol. 1, 286-291.

Walsh, K., Houston, W., and Houston, S. (1993). "Evaluation of in-place wetting using soil suction measurements," *J. Geot. Engr.*, ASCE, 119(5), 862-873.

Wheeler, S. and Sivakumar, V. (1992). "Critical state concepts for unsaturated soil," *Proc of the 7th Intl. Conf. on Expansive Clays*, Dallas, TX, August 3-5, Vol. 1, 167-172.

OSMOTIC SUCTION AS A VALID STRESS STATE VARIABLE IN UNSATURATED SOIL MECHANICS

by D.J. Miller[1] and J.D. Nelson[2]

ABSTRACT The hypothesis of the research being presented is that the osmotic suction of soil pore fluid is an independent, valid stress state variable for unsaturated soils. A review of concepts from interfacial thermodynamics and micromechanical equilibrium in colloidal systems is presented to illustrate the fundamental differences between matric and osmotic stress state variables. Ongoing experimental work described in more detail in a previous paper is oriented towards defining the roles of matric, osmotic, and total suction in the stress state and volume change behavior of unsaturated soils (Miller and Nelson, 1992). Testing is being conducted in two phases to determine: (1) matric and total suction moisture characteristics at various salt concentrations, and (2) unsaturated consolidation behavior in response to independently controlled changes in matric and osmotic suction. Preliminary results of the first phase of testing are discussed.

1. INTRODUCTION

Osmotic suction typically is neglected in a practical definition of the stress state of an unsaturated soil, with no profound consequences. Krahn and Fredlund (1971) concluded that for water content changes within the normal range of water contents encountered in most geotechnical engineering applications, the change in total suction is due primarily to changes in matric suction. But in situations where significant changes in pore fluid salt concentration will occur, there may be an associated change in osmotic suction that cannot be neglected.

A "true" effective stress equation for saturated soils has been widely disseminated in the soil mechanics literature in attempting to incorporate physico-chemical components in the definition of the stress state variable. The "true" effective stress" equation for saturated soils has been represented as (Lambe, 1960; Balasubramaniam, 1972; Chattapadhyay, 1972; Barbour and Fredlund, 1989)

---

[1] Instructor and PhD Candidate, Colorado State University

[2]Professor, Colorado State University, Civil Engineering Department, Fort Collins, CO 80523

$$\sigma'_{ij} = \sigma_{ij} - \delta_{ij}[u_f - (R-A)] \tag{1}$$

where, σ'_{ij} and σ_{ij} (i=j) are the effective and total normal stress components, σ_{ij} (i≠j) are the shear stresses, δ_{ij} is the Kronecker delta, u_f is the pore fluid pressure, and (R-A) is the net interparticle repulsive - attractive stress. The combination of (σ-u_f) and (R-A) in the normal stress terms of this matrix as a valid, single stress state variable has not been experimentally verified. We propose that it is not valid to incorporate parameters describing changes in microscale physicochemical stresses with macroscale mechanistic parameters in a single stress state variable. We hypothesize that the osmotic suction of the pore fluid - a macroscopic parameter - suitably characterizes stresses associated with changes in soil solution chemistry. Furthermore, osmotic suction should be represented as an independent stress state variable in the same way that total stress and matric suction are considered independently.

Osmotic suction is difficult to quantify by independent measurement, and typically is not determined in soil mechanics applications. Matric suction is commonly measured by tensiometers or axis-translation techniques, and total suction is measurable by psychrometric methods. It will be shown in the following discussion that total suction is equal to the sum of matric and osmotic components. However in general, osmotic and matric suction are not additive in terms of their effects on soil behavior. Therefore, total suction should not be used in place of matric suction when osmotic suction changes become important. The discussion presented in the following sections elaborates on the fundamental natures of total, matric, and osmotic suction.

2. INTERFACIAL THERMODYNAMICS AND EQUILIBRIUM

The interface between soil water solution and air is actually a surface between two immiscible phases. The physical properties of this interface are profoundly different from those of the bulk phases on either side. The most important characteristic of the interface, in relation to stresses in soils, is surface tension. Surface tension, T_s, is the coefficient of proportionality relating the work dw needed to change the surface area of an interface by an infinitesimal amount dA: dw = T_sdA. Its dimensions are energy/area (J/m^2 or N/m).

Capillary pressure is a direct result of surface tension. Consider, for example, the capillary rise of water in a small tube. Mechanical equilibrium between the air and water phases in the capillary tube is given by the Laplace Equation

$$p(g) - p(l) = \frac{2T_s}{r} \tag{2}$$

where, p(g) = air pressure, p(l) = water pressure, [p(g)-p(l)] = capillary pressure, and r = radius of the capillary tube (curvature of the meniscus at the air/water interface).

If instead of pure water a solution is used, T_s will be different and the capillary pressure will be affected. However, only constituents in the interfacial layer have a significant effect on T_s. For example, surfactants (substances that tend to accumulate in the interfacial layer) can have a significant impact on T_s and, consequently, on capillary pressure. Inorganic salts have the opposite effect as surfactants - they tend to have smaller concentrations in the interfacial phase than in the bulk solution. Therefore, the concentration of salt in the bulk solution must be very large to produce a significant change in T_s. Most inorganic salts produce a negligible increase in T_s in dilute water solutions in contact with air (Corey, 1977).

In soils, of course, the capillary geometry is much more complex than that of a single, symmetrical tube. At any given equilibrium water content, the *matric suction* represents an "equivalent capillary pressure" that is representative of the soil's pore size distribution. If it is assumed that surface tension is relatively unaffected by salt concentration (at least for dilute aqueous salt solutions), the matric moisture characteristic for a given soil matrix should not be significantly affected by changes in salt concentration in the soil solution.

Although inorganic salts do not alter matric suction to a significant degree, total suction, as measured by psychrometric techniques, is highly dependent on the concentration of salts in the pore water. Consider the capillary tube described previously, but containing a dilute salt solution instead of pure water. This is essentially a thermodynamic system comprising a simple liquid mixture of solvent (A) (water) and nonvolatile solute (B) (dissolved salt) contained in a capillary tube of radius r. It will be shown that, for this simple system, thermodynamic equilibrium between the solution and its vapor can be expressed by a single equation which lumps together the effects of capillarity and fluid chemistry.

Mechanical equilibrium is given by the Laplace equation (Eq.2), which is assumed to be unaltered by the chemical composition in the bulk liquid as discussed previously. Physicochemical equilibrium is represented by

$$\mu(g) = \mu(l) \tag{3}$$

where, $\mu(g)$ and $\mu(l)$ are the chemical potentials in the vapor and liquid phases, respectively. If the pressure is increased in the liquid phase, the chemical potential of the liquid changes by

$$d\mu(l) = V_m dp(l) \tag{4}$$

where, V_m = molar volume of liquid. Perfect gas behavior is assumed for the vapor phase

$$d\mu(g) = V_g dp(g) = \frac{RTdp(g)}{p(g)} \qquad (5)$$

where, R = gas constant per mole, and T = temperature (°K). Equating the changes in chemical potential gives

$$\frac{RT\ dp(g)}{p(g)} = V_m\ dp(l) \qquad (6)$$

In the case of pure water, when there is no additional pressure acting on the liquid, p(l) and p(g) are both equal to the normal vapor pressure $p(g)^*$. If the interface is curved towards the liquid, there is an additional pressure on the liquid given by the Laplace equation, such that $p(l) = p(g)^* + (2T_s/r)$. Thus for pure water in a capillary tube, Eq. 6 can be integrated as follows

$$RT \int_{p(g)^*}^{p(g)} \frac{dp(g)}{p(g)} = \int_{p(g)^*}^{p(g)^* + \frac{2T_s}{r}} V_m dp(l) \qquad (7)$$

It is assumed that the molar volume of the liquid is constant over the small range of pressures involved such that the molar volume can be taken outside the integral. On integration, the result is

$$\ln\frac{p(g)}{p(g)^*} = \frac{2T_s}{rRT} V_m \qquad (8a)$$

This is the Kelvin equation, which can be rewritten to obtain the vapor pressure over a curved surface of pure solvent

$$p(g) = p(g)^*\ e^{\frac{2T_s V_m}{RTr}} \qquad (8b)$$

To account for the presence of the solute, Eq. 3 must be extended to apply to the solvent A only: $\mu_A(g) = \mu_A(l)$. It can be shown that the effect of the curvature on the partial pressure of the solvent in the mixture is similar to that for the pure solvent liquid (Defay and Prigogine, 1966). The equation for vapor pressure reduction above the capillary surface of a solution is

$$\ln\frac{p_A(g)}{p_{A_o}(g)} = (\frac{2T_s}{rRT})V_A(l) \qquad (9)$$

where the subscript $_o$ denotes $1/r = 0$ (flat surface), $p_{Ao}(g)$ = partial pressure of solution with a flat surface, and V_A = partial molar volume of solvent A. Thus, the effect of the curvature on the partial pressure of the solvent in the mixture is similar to that for the pure solvent liquid (Kelvin equation, Eq. 8a). The partial pressure for an ideal solution with a flat surface is given by Raoult's law: $p_{Ao} = x_A p_A^*$, where x_A = mole fraction of A and p_A^* = vapor pressure of pure A. [Note: The notation (g) and (l) will be dropped from now on, since it is clear that pressures refer to the vapor phase and molar volumes to the liquid phase.] For dilute solutions $\ln x_A = \ln (1-x_B) \approx -x_B$, and V_A may be replaced by the molar volume of pure solvent V_m. Thus we arrive at an expression that relates the partial pressure of the solvent to the curvature and composition of the liquid

$$\ln\frac{p_A}{p_A^*} = \frac{2T_s}{rRT}V_m - x_B \tag{10a}$$

which can be rewritten in terms of the vapor pressure as

$$p_A = p_A^* \, e^{\frac{2T_sV_m}{rRT} - x_B} \tag{10b}$$

If Eq. 10b is compared to Eq. 8b, it is apparent that the effect of the nonvolatile solute is to reduce the vapor pressure over the capillary surface of the solution in comparison with the vapor pressure above a capillary surface of pure solvent with the same radius and surface tension (since for a nonvolatile solute $p(g) = p_A$ and $p(g)^* = p_A^*$). The ratio p_A/p_A^* is known as the relative humidity if the solvent is water.

Psychrometric methods for measuring total suction (e.g., filter paper tests and thermocouple psychrometers) provide an indirect method for measuring the relative humidity as it is controlled by a sample of soil. Total suction h is a measure of the reduction in chemical potential of the liquid solvent as a result of capillary pressure, and the presence of the solute. The general form for the chemical potential of a real (or ideal) solvent (or solute) is

$$\mu_A(1) = \mu_A^*(1) + RT \ln\frac{p_A}{p_A^*} \tag{11}$$

The reduction is from $\mu_A^*(l)$ to $\mu_A^*(l) + RT\ln(x_A)$ (since $x_A < 1$). This difference is known as the Gibbs free energy ($\Delta G = \mu_A - \mu_A^*$). It is customary in soil science to multiply ΔG by (ρ_f/M) to convert to pressure units, where $\rho_f =$ density of salt solution and $M =$ molecular weight of the solution

$$h = \frac{\rho_f}{M}[\mu_A(1)-\mu_A^*(1)] = \frac{\rho_f RT}{M}\ln\frac{p_A}{p_A^*} \tag{12}$$

An expression for the total suction in soil can be obtained by substituting the expression for relative humidity for a capillary solution (Eq. 10a) into Eq. 12, which gives

$$h = \frac{\rho_f}{M}(-\frac{2T_s}{r}V_m - RTx_B) = -\frac{\rho_f V_m}{M}(\frac{2T_s}{r} + RTC) \tag{13}$$

The last quantity on the right hand side of Eq. 13 (RTC) is the osmotic pressure (osmotic suction) of the solution in accordance with the van't Hoff equation

$$\pi = RTC \qquad (14)$$

where, π is in units of kPa, if R is in units of 8.32 L kPa/ °K mol), T in °K, and C = molar concentration of solute in mol/L = $x_B V_m$. Thus, total suction is equal to the sum of capillary $(2T_s/r)$ and osmotic (RTC) pressures.

In a soil with a given gradation, void ratio, and water content (i.e. given "effective" capillary radius) the Laplace equation (Eq. 2) implies that matric suction will be the same for two samples with different pore fluid compositions, assuming that composition has a negligible impact on surface tension. However, the total suction, which represents the combined effects of capillarity and vapor pressure reduction due to the dissolved salts in the bulk solution will be different for the two samples, as shown by Eq. 13. These phenomena were observed in the preliminary Phase I test results, described below.

3. MACROSCALE VERSUS MICROSCALE CONSIDERATIONS

Why not use total suction as a single stress state variable, since it represents the combined magnitudes of matric and osmotic suctions? In fact, the related concept of a "total soil water potential" was considered for many years in the field of soil science as the fundamental driving force governing the flow of water through soils. But Corey and Kemper (1961) demonstrated that there is no single potential that is a function of the state of the *water* only whose gradient will always indicate the direction of net transport of water, or which, if constant in all parts of the soil system, will ensure that equilibrium exists.

Total suction *is* a measure of the combined mechanical and physicochemical equilibrium state of the soil water solution. But it does not follow that mechanical (matric) and physicochemical (osmotic) *stresses* are additive, any more than the total soil water potential governs the net transport of water in soils. The basic problem with defining a single, "total potential" to define equilibrium has to do with maintaining the distinction between microscale and macroscale representative elementary volumes (REVs) associated with each component of total potential. Matric potential is defined with respect to a macrosopic element comprised of soil particles, soil solution, and air. Osmotic potential is defined for elements of the chemical species H_2O as a component of the soil solution. It is not valid to sum forces relevant to different kinds of REVs. The only time water movement, for example, will be governed by a gradient in total potential is when the soil acts as a true *semipermeable* membrane, in which case all species of the soil solution except water are prevented from being transported. In the general case, where all species are free to move, or are only partially restrained, it is necessary to define the magnitude of each potential gradient separately and determine the coefficients relating the flux to these separate potential gradients independently, as they depend on different properties of the soil medium. Corey and Klute (1985), developed a general theory of water transport and equilibrium, emphasizing the distinction between microscopic and macroscopic viewpoints. Barbour (1987) and Barbour and Fredlund (1989) considered this in their developments of osmotic flow and consolidation.

It is proposed that the distinction between microscale and macroscale should be maintained in terms of stress state variables. Total (or effective) stress and matric suction are mechanical stresses with respect to a macroscopic REV. Fredlund (1973) set up a model using principles of multiphase continuum mechanics for a four phase soil system comprising solid, liquid, gas, and the liquid/gas interfacial surface which Fredlund referred to as the "contractile skin". Fredlund and Morgenstern (1977) showed that the stresses σ = total stress, u_w = pore water pressure, and u_a = pore air pressure, in an REV of the four phase system could be combined in a number of different ways to form valid, independent stress state variables for unsaturated soils. Two of these variables are needed for a complete definition of the stress state in unsaturated soils, when changes in pore solution chemistry can be neglected. They recommended the combination

$$[\sigma_{ij} - \delta_{ij}u_a] \quad \text{and} \quad [\delta_{ij}(u_a - u_w)] \tag{15}$$

The first tensor is the total stress state variable and the second is the matric suction stress state variable. The validity of these stress state variables has been demonstrated both experimentally (Fredlund, 1973) and theoretically on the basis of thermodynamic principles (Edgar, 1983), *for soils in which pore fluid chemistry is unchanging*.

Osmotic suction reflects the pore fluid chemistry, but impacts interparticle stresses on a fundamentally different scale and by fundamentally different mechanisms. Most of the research done in this area has been focused on the behavior of colloidal suspensions of clays. In such systems, microscale "double layer" theory is used with some success. According to microscale theory, the "thickness" of diffuse double layers, and the degree to which ions held within those regions are restricted from moving along with the bulk solution, depend primarily on the salt concentration of the bulk solution. On the microscale, interparticle repulsive forces in clays are attributed primarily to overlapping double layers, and the *osmotic pressure concept*. This concept, simply stated, says that the repulsive force between clay particles is given by the net osmotic pressure π_{net} between the bulk solution and the mid-plane between two parallel, charged particles. In a real soil system subjected to changes in pore fluid chemistry, changes in π of the bulk solution essentially will be equivalent to changes in π_{net}. For complete and detailed discussions of the osmotic pressure concept see, for example, Mitchell (1976), Adamson (1990), and Bolt (1956).

4. STRESS STATE VARIABLES AND VOLUME CHANGE CONSTITUTIVE RELATIONSHIPS

It is proposed that the change in soil stress associated with a change in chemistry of the soil solution is adequately quantified by the change in solution π. Although π has been introduced in the literature as being a somewhat inferior substitute for (R-A) as a stress state variable, it is the contention of these authors that π is actually a more valid candidate for that role. A macroscopic stress state variable must be *independent* of the physical properties of the body for which the stress state is being defined. Incorporation of the microscale forces into a stress state variable for soils violates this requirement because (R-A) is highly dependent soil and solution properties. π is a macroscopic parameter, and is independent of the physical properties of the soil.

It is hypothesized that the following independent variables can be utilized to fully define the stress state of soils:

$$[\sigma - \delta_{ij}u_a] \quad = \quad \text{total stress state variable}$$
$$\delta_{ij}[u_a - u_w] \quad = \quad \text{matric suction stress state variable} \qquad (16)$$
$$\delta_{ij}\pi \quad = \quad \text{osmotic suction stress state variable}$$

The constitutive relationships that relate to volume changes in a soil can be written in terms of void ratio (e) and degree of saturation (S_r). Linear relationships for these in terms of the three stress state variables given in Eq. 16 are

$$de = a_t \, d(\sigma-u_a) + a_m \, d(u_a - u_w) + a_\pi \, d(\pi)$$
$$(17)$$
$$dS_r = b_t \, d(\sigma-u_a) + b_m \, d(u_a - u_w) + b_\pi \, d(\pi)$$

The coefficients a_t, a_m, and a_π represent the coefficients of compressibility with respect to changes in total stress, matric suction and osmotic suction stress state variables, respectively. The coefficients b_t, b_m, and b_π are the slopes of the saturation vs. respective stress state variable plots. Alternative logarithmic forms of these equations may be more applicable over a broader range of stress.

5. LABORATORY TESTING PROGRAMS

A comprehensive laboratory investigation has been initiated to assess the validity of the proposed stress state variables and constitutive relationships. The primary goal of the testing program is to determine whether the osmotic suction of a soil can serve as a valid, independent stress state variable in a macroscale, mechanistic model of unsaturated (and saturated) soil stress-strain behavior. The testing program, and preliminary results of the first phase of testing were presented in Miller and Nelson (1992). A brief summary and discussion is provided here.

The testing program has two phases: (1) determination of suction-moisture-volume characteristics for various concentrations of NaCl in the pore solution of a compacted clayey soil, and (2) measurement of soil compressibility under controlled changes of the proposed stress state variables.

The initial experiments in Phase I entailed measurements of both total and matric suction changes with changes in water content. Matric suction moisture characteristics (drying curves) were determined using a conventional pressure plate device. Total suction moisture characteristics were determined by the filter paper technique. Comparison between the total and matric suction moisture characteristics will provide some insight regarding the role of osmotic suction on moisture retention.

Matric and total moisture characteristics are presented in Figures 1 and 2, respectively. The drying curves are plotted in terms of gravimetric water content. The following observations are made with regard to these data:

• In terms of matric suction (Fig. 1), there is no apparent distinction between moisture retention relationships for samples prepared with distilled water, tap water, and NaCl solution.

• In terms of total suction (Fig. 2), there are significant differences between drying curves for samples compacted with distilled water and those prepared with NaCl solution. The total suction moisture characteristic for samples compacted with 2% by dry weight NaCl lies well above that for distilled water samples at high moisture contents. The curves converge as moisture decreases and intersect at about 3% (g/g) moisture.

Volume changes during drying were obtained by periodically weighing the specimens and measuring their dimensions with precision calipers. Final oven dry weight and volume then were determined. Results of the shrinkage measurements are shown in Figure 3. The difference between the distilled water and NaCl samples is evident. At any given water content on the shrinkage curves, the specimen volumes are smaller for samples prepared with NaCl.

The shrinkage curves shown in Fig. 3 indicate that the added NaCl had a measurable impact on the strain state (volume) of the soil at any given water content. However, Fig. 1 shows that the matric suction moisture characteristic is essentially unaffected by the salt concentration over a range of water contents between approximately 14 to 20% (the measurable range of moisture change on the pressure plate). The observed difference in the relationship between specimen volume and water content (Fig. 3) indicates that there is an additional stress acting on the soil that is not accounted for by the total stress and matric suction stress state variables only. This interpretation of the data is supported by the fact that, in terms of total suction, there is a significant difference between the NaCl and distilled water sample

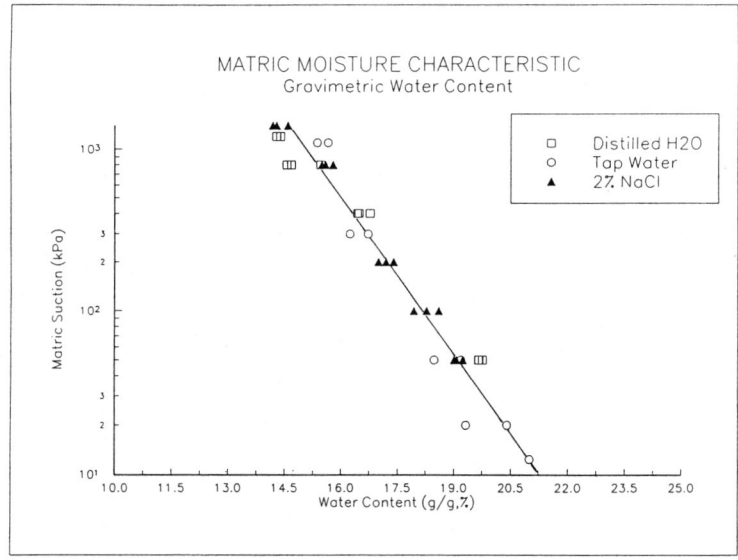

Figure 1

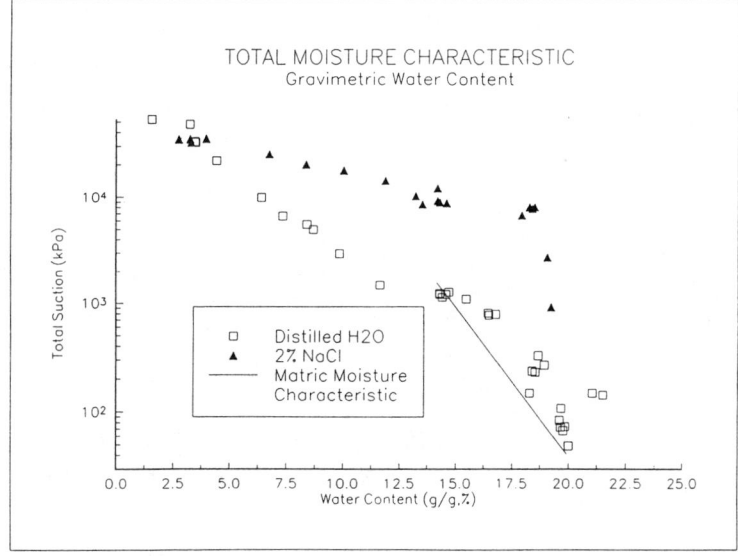

Figure 2

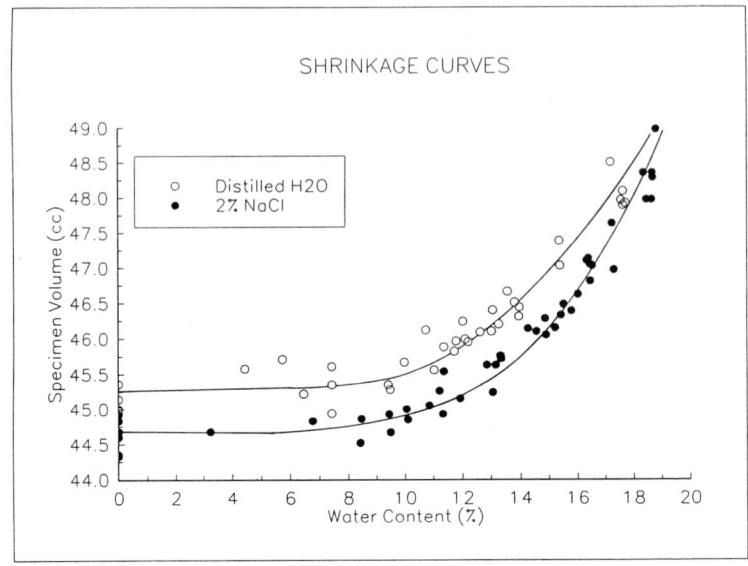

Figure 3

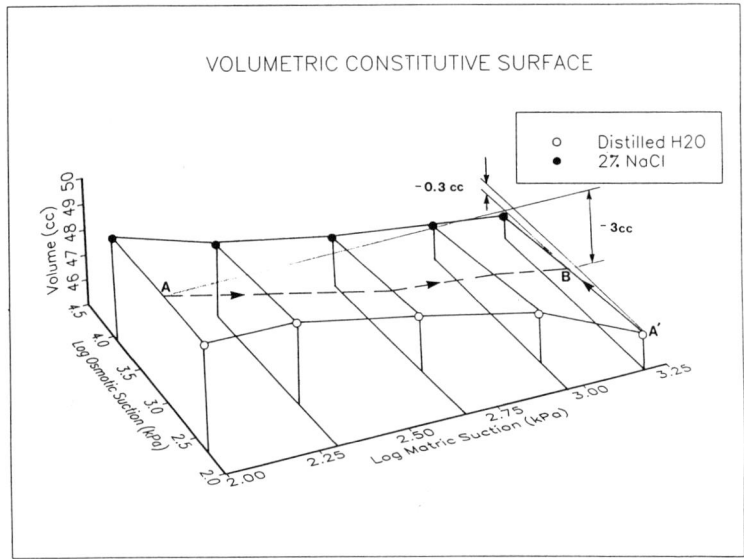

Figure 4

moisture characteristics as shown on Fig. 2. Since total suction comprises both capillary and osmotic mechanisms, at any given capillary pressure, the total suction will be higher for salt solutions than for pure water. This explains why the total moisture characteristic for the salt-treated samples plots above that for the distilled water samples. If the volume difference is associated with an "osmotic stress", the total moisture characteristics verify that these stresses, in fact, do exist in the salty samples.

The data are plotted three dimensionally in Figure 4. Five different water contents were chosen between 14 and 19 percent, which corresponds with the the measurable matric suction range between approximately 12 bar to 1 bar, as determined by conventional pressure plate testing. Matric and total suction values corresponding to each water content were determined from the moisture characteristic plots (Figs. 1 and 2). Osmotic suction was calculated as the difference between total and matric suction at each water content. Specimen volumes were similarly interpreted from Figure 3. Specimen volume (cc) is plotted against log matric suction and log osmotic suction (kPa) in Figure 4.

Figure 4 illustrates that the soil is more compressible in response to changes in matric suction than in response to osmotic suction changes, in the water content range studied. The overall volume change from point A to point B in Fig. 4 is approximately -3 cc, corresponding to a matric suction change from approximately 100 to 1500 kPa. Volume change in the other direction showing an increase in π at constant (u_a-u_w) (from point A' to B), and corresponding to a change in osmotic suction from 100 to 1500 kPa, is only about -0.3 cc.

On the basis of these preliminary test results, it is hypothesized that osmotic suction constitutes an additional stress on the soil. The compressibility of the soil in response to changes in osmotic suction appears to be significantly less than the compressibility in response to changes in matric suction, for water content changes within the normal range of measurement of matric suction. Based on theoretical considerations elaborated on previously, it is proposed that osmotic suction be treated as an independent stress state variable. This hypothesis will be tested by conducting additional Phase I tests at different salt concentrations, and in Phase II of the laboratory testing program.

6. REFERENCES

Adamson, A.W. (1990). *Physical Chemistry of Surfaces*, Wiley, New York.

Balasubramanian, B.I. (1972). "Swelling of compaction shale." PhD Dissertation, Univ. Alberta, Edmonton, Alberta.

Barbour, S.L. (1987). "Osmotic flow and volume change in clay soils." PhD Dissertation, Univ. Saskatchewan, Saskatoon.

Barbour, S.L. and Fredlund, D.G. (1989). "Mechanisms of osmotic flow and volume change in clay soils." *Can. Geotech. J.*, 26, 551-562.

Bolt, G.H. (1956). "Physico-Chemical Analysis of the Compressibility of Pure Clays," *Geotechnique*, **6**, 86-93.

Chattopadhyay, P.K. (1972). "Residual shear strength of some pure clay minerals." PhD Dissertation, Univ. Alberta, Edmonton, Alberta.

Corey, A.T. (1977). *Mechanics of Heterogenous Fluids in Porous Media*, Water Resources Publications, Fort Collins, CO.

Corey, A.T. and Kemper, W.D. (1961). "Concept of Total Potential in Water and its Limitations," *Soil Science*, **91**(5): 209-302.

Corey, A.T. and Klute, A. (1985). "Application of the Potential Concept to Soil Water Equilibrium and Transport," *Soil Sci. Soc. Am. J.*, **49**(3): 3-11.

Defay, R. and Prigogine, I. (1966). *Surface Tension and Adsorption*, Wiley.

Edgar, T.V. (1983). "Moisture movement in nonisothermal deformable media." PhD Dissertation, Colorado State Univ., Fort Collins, CO.

Fredlund, D.G. (1973). "Volume change behavior of unsaturated soils." PhD Dissertation, Univ. Alberta, Edmonton, Alberta.

Fredlund, D.G. and Morgenstern, N.R. (1977). "Stress state variables for unsaturated soils." *J. Geotechnical Engrg. Div.*, ACSE, 103(5), 447-466.

Krahn, J. and Fredlund, D.G. (1971). "On Total, Matric and Osmotic Suction," *Soil Sci.* **114**(5), pp. 339-345.

Lambe, T.W. (1960). "A Mechanistic Picture of the Shear Strength of Clay," *Proc. ASCE Research Conf. Shear Strength of Cohesive Soils*, p. 437.

Miller, D.J. and Nelson, J.D. (1992). "Osmotic Suction as a Valid Stress State Variable in Unsaturated Soils," *7th Intern. Conf. Expansive Soils*, pp. 179-184.

Mitchell, J.K. (1976). *Fundamentals of Soil Behavior*, Wiley, New York.

ASPECTS OF THE BEHAVIOR OF CLAYS ON DRYING

F.A.M.Marinho[1] and R.J.Chandler[1]

Abstract

The drying behavior of compacted London Clay mixed with varying proportions of fine sand is examined, measuring the changes of sample volume, water content and suction. Particular attention is paid to the relationships between water content and suction, and to the suctions at which saturated or near-saturated samples commence to desaturate.

The relationships between water content and the logarithm of suction are found to be linear, independent of initial water content, confirming earlier work. These relationships vary systematically with the liquid limit of the soil, enabling a general relationship to be proposed for compacted soils. Desaturation of near-saturated soils commences at suctions varying between 7 MPa for pure London Clay with liquid limit of 77%, and 65 kPa for a London Clay/sand mixture having a liquid limit of 24%.

Introduction

A number of authors have, over the years, studied the behavior of soil on drying (*e.g.* Poulovassilis (1970), Fleareu *et al.* (1992)). Such studies are of interest for a number of reasons; for example, there is need to relate drying behavior, in which progressive desaturation occurs, to the behavior of saturated clays during consolidation. The question as to when does an initially saturated soil subjected to drying first desaturate, and at what suction, is closely related to the question as to the maximum suction a saturated

---

[1]   Department of Civil Engineering, Imperial College of Science, Technology and Medicine, London SW7 2BU, United Kingdom

sample can sustain when removed from the ground. It is also of interest to examine how the drying behavior varies with such factors as the plasticity of the soil. These factors are addressed in this paper.

Methods

A series of compacted samples was prepared using London Clay, and mixtures of London Clay and fine sand. In all, five different mixtures were used, ranging from 100% London Clay to 30% clay with 70% sand. Between five and nine compacted samples were prepared for each mixture, each at a different water content, a total of 39 samples. The index properties, number of samples, and range of preparation water contents for each soil mixture are given in Table 1.

Clay/sand mixture	Liquid limit (%)	Plastic limit (%)	No. samples	Water content range (%)
100/0%	77	29	7	21.4 - 30.5
90/10%	69	24	9	14.4 - 29.8
70/30%	54	19	9	11.8 - 27.1
50/50%	40	17	9	12.6 - 20.4
30/70%	24	18	5	8.5 - 16.1

Table 1. Summary of testing programme.

The samples were all compacted in a 38 mm diameter cylindrical mould, using a sliding weight of 1000 g mass falling through 305 mm. Each sample was compacted in 10 layers, with six blows per layer. This corresponds to a compactive effort, or work done per volume, of 2100 kJ/m^3, which compares to 596 kJ/m^3 for the AAHSHO/BS Standard Compaction Test and 2682 kJ/m^3 for the AAHSHO/BS Heavy Compaction Test. When compacted the samples each had a length of about 75 mm. Subsequent references to optimum water content relate to this particular compaction procedure, which, as can be seen, is quite close to the AAHSHO/BS Heavy Compaction Test.

Some practical difficulties were encountered with the 100/0% clay/sand mixtures, partly when removing samples from the compaction mould, but also because slight cracking occurred between layers on drying. Mention is made in the text where these problems appear to have affected the results. In addition, two specimens of the 50/50% clay/sand mixture were unintentionally more heavily compacted than described above. These are the two points that lie above the general range of values plotted in Figure 1. In spite of this, the data yielded by these two samples is consistent with other results, and they are reported herein.

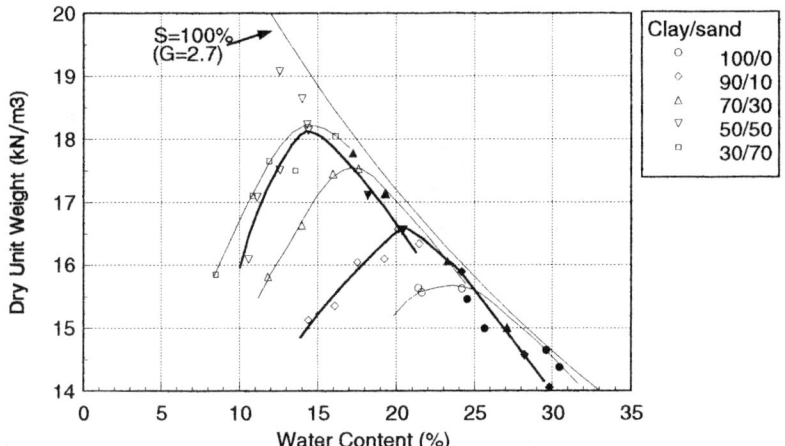

Figure 1. Compaction curves for the various soil mixtures. The solid symbols are those samples that exhibit "near-saturated" behavior on drying.

Following compaction, each sample was allowed to dry naturally in the laboratory atmosphere, the drying process being interrupted for the suction to be measured. Suction measurements were made with the filter paper technique, using Whatman's Nº. 42 paper, following the test procedures of Chandler & Gutierrez (1986) and Chandler et al. (1992). The correlations between filter paper water content and suction used here are presented in the latter reference. Two separate filter papers were placed in contact with each sample, the average water content of the two papers being used to obtain the suction in the sample. At higher soil water contents this technique will measure matrix suction, and, as the sample dries, there will be a gradual transition to a measurement of total suction. As will be seen later, since the degree of saturation remained high for many of the samples, it is believed that the large majority of the measurements made were of matrix suction.

A period of seven days was allowed at each suction measurement for the filter papers to achieve equilibrium with the soil sample, the sample being sealed from the atmosphere by being wrapped in two layers of plastic 'cling-film' and then placed in a sealed polythene bag. Since up to six suction measurements were made during drying, the drying test on each individual sample lasted for up to six weeks.

Each sample was weighed (to 0.05 gm) both at each suction measurement and at intermediate stages during drying; at the same time, the dimensions of each specimen were carefully measured using vernier callipers reading to 0.01 mm. These measurements enabled the overall volume of each sample to be calculated, and, knowing the specific gravity (G) and the mass of the soil grains, the void ratio, water content and degree of saturation (S%) can then also be calculated. The specific gravity of the London Clay was 2.75, that of the sand 2.65, so that the soil mixtures had average specific gravities varying between 2.75 and 2.67. For simplicity, the idealised relationships for S = 100% shown on the various figures have all been calculated using a specific gravity of 2.70.

While these techniques inevitably are somewhat imprecise, they have the advantage of relative simplicity, enabling reasonable numbers of tests to be carried out. Thus the results can be used to establish trends of behavior which may later be confirmed, if necessary, by more precise methods.

Results

As will be seen from Figure 1, the preparation water contents span either side of the optimum water content of each of the soil mixtures. Each sample, therefore, commenced drying from a different dry unit weight and different degree of saturation. With the 'pure' London Clay samples this resulted in a range of initial degrees of saturation between 80% and 100%; as the sand content increased the initial degree of saturation fell, so that with the 30/70 clay/sand mixture the initial degrees of saturation ranged between 35% and 95%.

The results of the drying tests are plotted as the sample volume per 100 g mass of dry soil, following Haines (1923). This idealisation shows the behavior of an initially saturated sample which commences to desaturate as it approaches the shrinkage limit, Figure 2.

Experimental drying curves are plotted in this manner in Figures 3, 4 and 5. These three figures show the results for clay/sand mixtures of 100/0, 70/30 and 30/70%; the results for the other two mixtures (90/10 and 50/50% clay/sand) are in all respects intermediate between those shown, and are omitted for reasons of space. Figures 3, 4 and 5 also show the theoretical saturation line assuming G = 2.70.

It will be seen that the drying relationships fall in two types, depending on the initial water content of the sample. Where the preparation water content lies on the wet side of optimum, the drying relationship initially lies on, or parallel and close to, the S = 100% line. At lower preparation water contents the drying curve may show an initial linear portion, though this diverges from the S = 100% line. This behavior is particularly obvious at the lowest preparation water contents. Drying parallel to the S = 100% line implies that the volume of air remains constant, while a diverging drying line is consistent with an increasing volume of air. The point where the drying curve departs from the S = 100% line will be referred to as the point of "general air entry". With initially unsaturated samples the point where general air entry occurs is best determined from a plot of S against water content, as discussed later. Samples that show drying curves along or parallel to the saturation line are referred to as "near-saturated"; it seems unlikely that compacted samples are truly saturated. These near-saturated samples are shown as solid symbols in Figure 1, where it will be seen that all lie close to the saturation line.

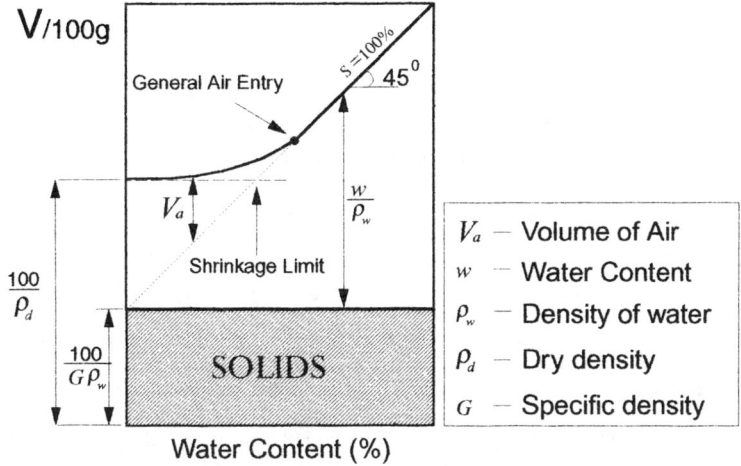

Figure 2. Idealised drying behavior of 100g mass of dry soil.

The experimental results can also be plotted against the interrelated parameters of void ratio, degree of saturation, water content and suction as shown in Figures 6, 7 and 8. Again, and for similar reasons as before, only the data relating to clay/sand mixtures of 100/0, 70/30 and 30/70% are shown. The void ratio versus water content relationships shown in Figures 6-8 are effectively the same as those in Figures 3-5. The plot of saturation versus water content is valuable in demonstrating the point (and the rate) at which desaturation occurred on drying.

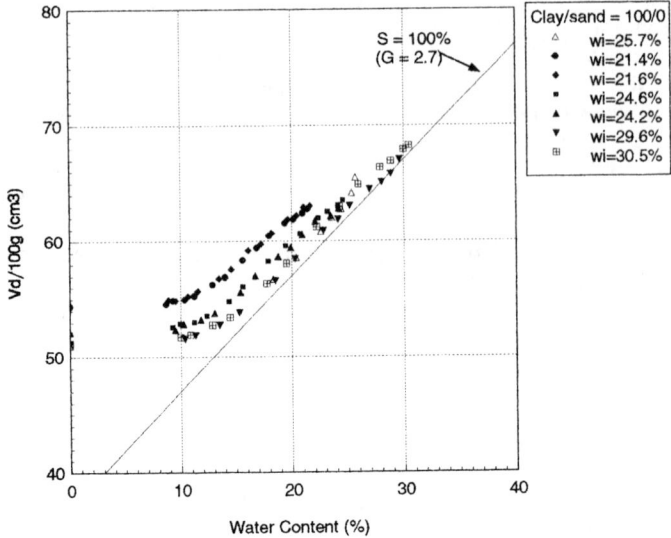

Figure 3. Drying behavior of 100/0% clay/sand mixture; V_d volume per 100gms of soil solids.

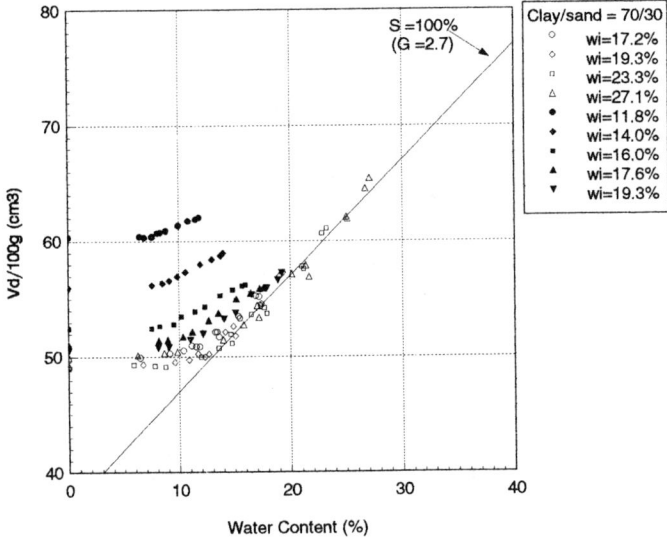

Figure 4. Drying behavior of 70/30% clay/sand mixture; V_d volume per 100gms of soil solids.

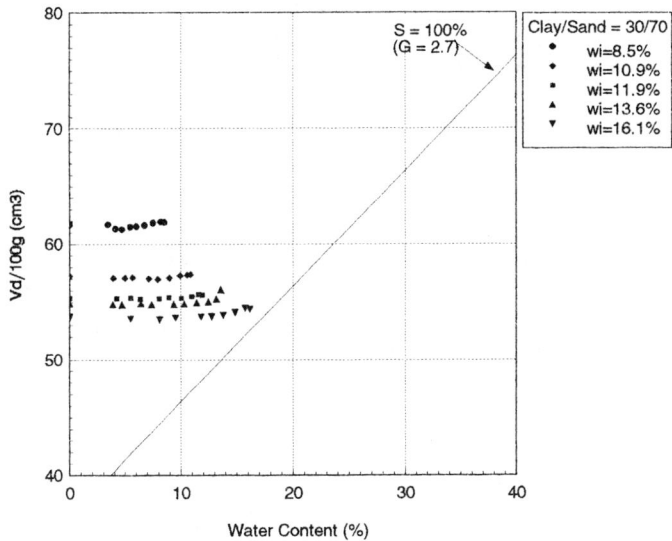

Figure 5. Drying behavior of 30/70% clay/sand mixture; V_d volume per 100gms of soil solids.

Discussion

Interesting comparisons can be made in Figures 6-8 between the relationships of void ratio versus suction and water content versus suction. With 100/0% clay/sand material the relation between void ratio and suction appears to be largely independent of the preparation water content. As the proportion of sand increases, it becomes clear that there is not a unique relation between void ratio and suction, and that significantly different relationships must be followed by those samples which had lower preparation water contents. These various relations presumably converge at high stresses, as suggested by the dashed lines in Figures 7 and 8. In contrast, the relationships between water content and suction do appear to be unique, independent of the preparation water content. Moreover, the relationships are linear with respect to the logarithm of suction. Similar conclusions have been reached by other workers; see, for example, the review of such data by Ho *et al.* 1992.

SUCTION CAPACITY The relationships between water content and suction for all five soil mixtures are combined in one diagram in Figure 9. It will be seen that there is a systematic relationship between the data, the more plastic mixtures exhibiting higher water contents at a given suction. The plotted

84 UNSATURATED SOILS

lines are the regression lines for all the data points for each soil mixture, apart from the 100/0% clay/sand mixture for which the relationship shown has been plotted noting the trend of the relationships found for the other soil mixtures. The greater scatter of the data for the 100/0% mixture is probably related to the practical difficulties encountered in compacting this soil.

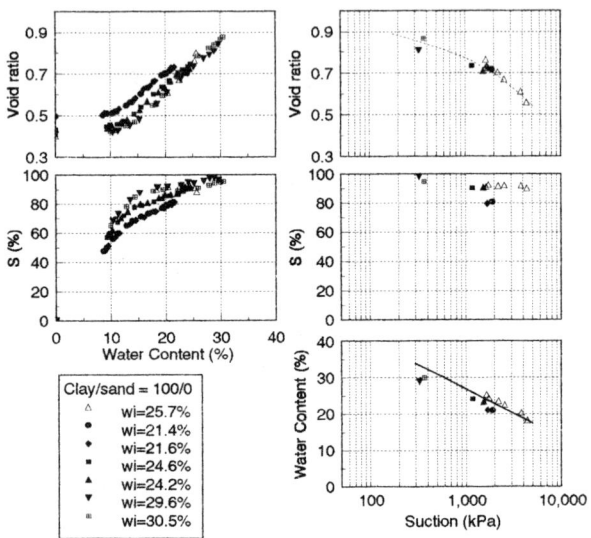

Figure 6. Drying behavior of 100/0% clay/sand mixture plotted against common axes of water content and suction; S degree of saturation.

The slope of the lines varies with the plasticity of the soil mixture. This slope can be described as the reduction in water content (expressed as percent) over one logarithmic cycle. This will be termed the "Suction Capacity", noting that Richards (1931) defined a very similar "capillary capacity". Where possible, the Suction Capacity is determined over the range of suction from 100 kPa to 1000 kPa. Thus the Suction Capacity can be compared with the Compression Index used for the one-dimensional consolidation behavior of saturated soils, which is similarly defined over one log-cycle. Other workers (e.g. Alonso et al. (1987), Ho et al. (1992)) have also made use of the water content-suction relationship.

Plotting the Suction Capacity against liquid limit yields a direct relationship, Figure 10, to which may be added data from Ho et al., (1992), also for dynamically compacted samples, which strongly support the results of the present study.

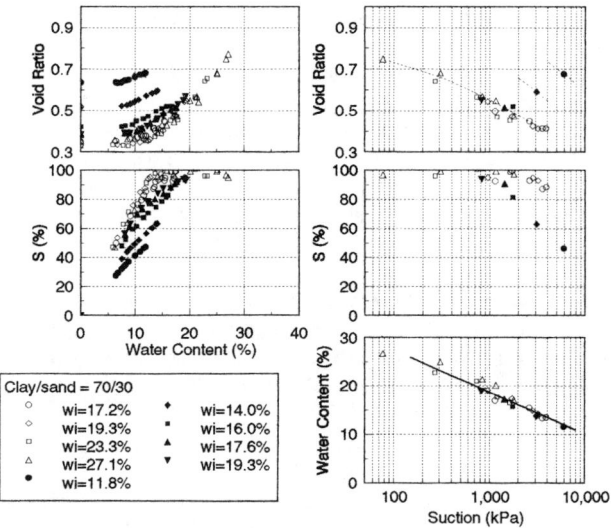

Figure 7. Drying behavior of 70/30% clay/sand mixture plotted against common axes of water content and suction; S degree of saturation.

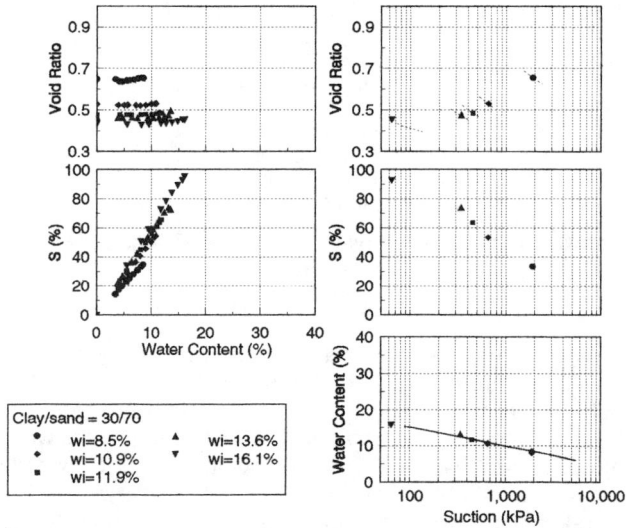

Figure 8. Drying behavior of 100/0% clay/sand mixture plotted against common axes of water content and suction; S degree of saturation.

The relationship between suction capacity and liquid limit in Figure 10 holds only for compacted soils. For comparison, in situ consolidation data for saturated normally consolidated clays from Burland (1990) are plotted as Suction Capacity against liquid limit in Figure 11. In doing this, Suction Capacity has been calculated with reference to mean effective stress $(= (\sigma_1' + 2\sigma_3')/3$, assuming $K_o = 1 - \sin\phi')$. As will be seen, the compacted soil data fall below that for the normally consolidated soils. Limited data from drying tests by the authors on undisturbed heavily overconsolidated clays show lower Suction Capacities than the compacted samples.

The data plotted in Figure 9 can be further normalised with respect to Suction Capacity, as shown in Figure 12, where a near unique linear relation is shown to exist between water content normalised by Suction Capacity, and the logarithm of suction. Again, it must be emphasised that this particular relationship can only be expected to hold for compacted soils.

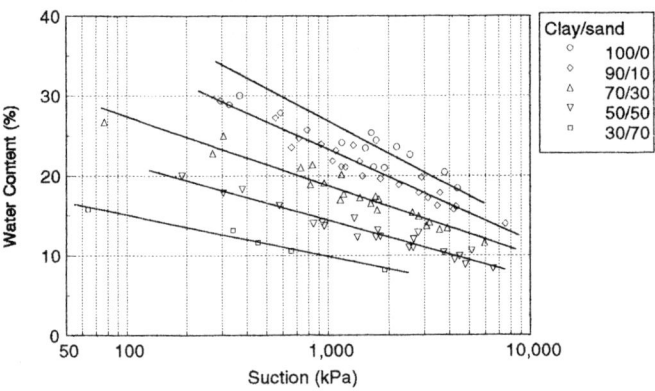

Figure 9. Relationships between water content and suction for all five soil mixtures.

SATURATED BEHAVIOR   It has been noted that the various drying curves either initially follow the saturation line ("near-saturated behavior") or they immediately diverge from it, depending on the initial water content. Typically the near-saturated samples were compacted wet of optimum, with an initial degree of saturation of 95-100%. Samples compacted at or dry of optimum have an initial degree of saturation less than 95%. The drying behavior in terms of the relation between degree of saturation and water content is bilinear, the point where the change in gradient occurs being defined, as shown in Figure 13, as the point of general air entry.

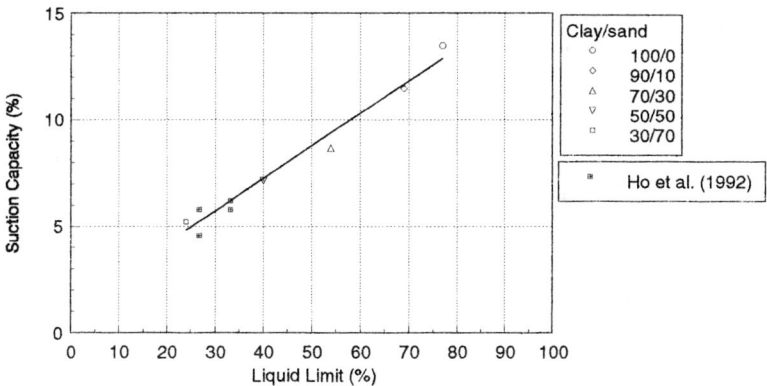

Figure 10. Suction capacity versus liquid limit for compacted soils.

The importance of the point of general air entry is that it is provides a useful means of assessing the range of saturated behavior of initially near-saturated samples. These samples are found to reach the point of general air entry at degrees of saturation in excess of 90%, whereas those samples which initially had lower degrees of saturation show a more rapid reduction in the degree of saturation in relation to changing water content, and show a much lower degree of saturation at general air entry, as illustrated in Figure 13. In the present study the point of general air entry for samples compacted near or dry of optimum occurred at degrees of saturation in the range from 85% to as low as 50%.

The suctions developed for near-saturated samples at the point of general air entry can be obtained from the data in Figures 6-8, and are shown in Figure 14 plotted on the relations between water content and suction. The 100/0% samples appear to be capable of sustaining suctions as high as 7 MPa before "desaturation", but the less plastic samples show progressively lower suctions at general air entry. The point of general air entry for near-saturated 30/70% material was as low as 65 kPa.

Conclusions

The study of the drying behavior of compacted London Clay, and mixtures of London Clay and fine sand, confirm previous work on this topic in the context of the development of suctions during drying. The preparation of samples at a range of saturations from close to 100% ("near-saturated"), to less than 40%, enabled the drying process to be compared over a wide range of initial conditions. For all the soil mixtures studied, the water content was found to be linearly related to the logarithm of suction, confirming findings by earlier workers. If the slope of the water content versus logarithm of suction relationship is defined as the Suction Capacity, C, then the present work confirms that there is a close relationship between C and the liquid limit of the soil. Indeed, the sample water contents may be normalised by C to define a unique relation between w/C and the logarithm of suction that should be of value for predictive purposes with compacted soils. Comparable relationships will need to be established for naturally occurring soils in situ.

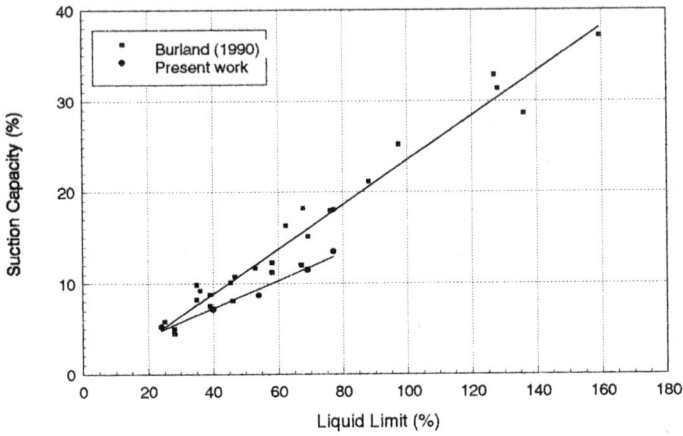

Figure 11. Suction Capacity for normally consolidated soils plotted against liquid limit, for comparison with data from figure 10.

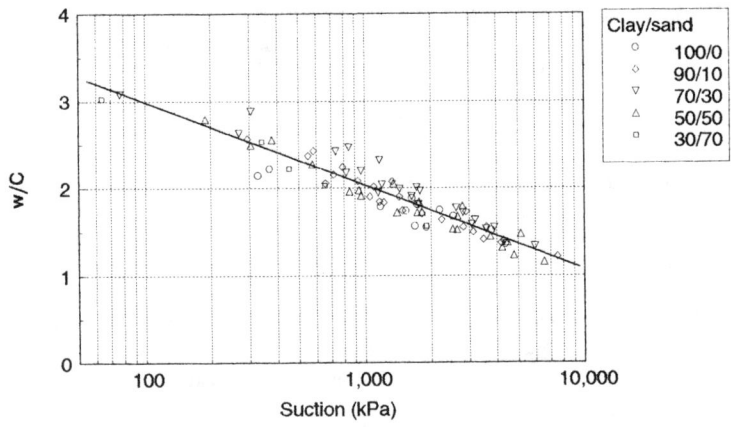

Figure 12. Relationships from figure 9, with water content normalised by suction capacity.

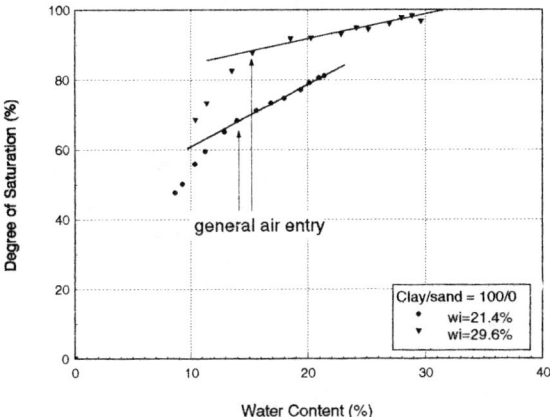

Figure 13. Determination of "general air entry".

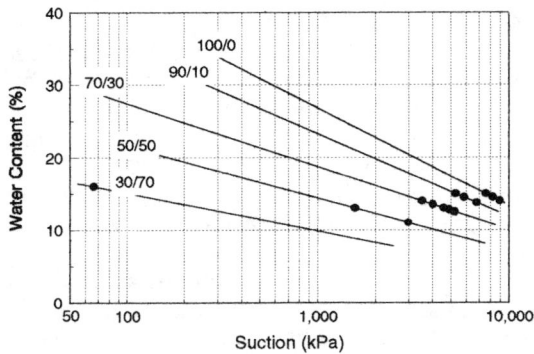

Figure 14. Suctions at general air entry for saturated or near-saturated samples. The relationships are from figure 9, with general air entry points determined as shown in figure 13.

Soils initially compacted with a degree of saturation of 95% or higher showed drying behavior in the manner anticipated for initially saturated soils. Samples with an initial degree of saturation greater than about 80% and with liquid limits of 40% or more showed a clear point of "general air entry". The point of general air entry is best defined by the relation between water content and degree of saturation, and shows as a clear change of gradient, even for samples where the initial degree of saturation is less than 95%. The suctions measured at the general air entry point for near-saturated samples provide an indication of the suctions that might be sustained, for example during sampling. These suctions range from 7 MPa for the pure London

Clay, to about 65 kPa for the 30/70% London Clay/ sand mixture, which had a liquid limit of 24%.

Acknowledgements

The first author is grateful for sponsorship by the Brazilian Government agency CAPES. The authors would like to thank Mr. Zoltan Elephanty for carrying out some of the laboratory work.

Appendix   References

Alonso, E.E., Gens, A. and Hight, D.W. (1987) General report-Specialproblem soils - 9th European Conference of Soil Mechanics andFoundation Engineering .**3**, 1087-1146.

Burland, J.B. (1990). On the compressibility and shear strength of natural clays. Géotechnique **40**, 329-378.

Chandler, R.J., Crilly, M.S. and Montgomery-Smith, G. (1992). A low-cost method of assessing clay desiccation for low-rise buildings. Proc. Instn. Civ. Engrs., Civ. Engng., **92**, 82-89.

Chandler, R.J. and Gutierrez, C.I. (1986). The filter-paper method of suction measurement. Géotechnique **36**, (2), 265-268.

Fleareu, J.M., Soemitro, R. and Taibis, S. (1992) Behavior of an expansive clay related to suction. 7th International Conference on Expansive Soil,- Dallas, Texas, **1**, 173-178.

Haines, W.B. (1923). The volume changes associated with variations of water content in soil. Journal of Agricultural Science, **13**, 296-310.

Ho, D.Y.F., Fredlund, D.G. and Rahardjo, H. (1992). Volume change indices during loading and unloading of an unsaturated soil. Canadian Geotechnical Journal, **29**, 195-207.

Poulovassilis, A. (1970). The hysteresis of pore water in granular porous bodies. Soil Science,.**109**, 5-12.

Richards, L.A. (1931). Capillary conduction of liquids in porous mediums. Journal of General and Applied Physics, American Physical Society,**1**, 318-333.

The Transistor Psychrometer

A New Instrument for Measuring Soil Suction

John A.Woodburn[1], James C. Holden[2] and Paul Peter[3]

Abstract

The transistor psychrometer is the latest instrument available for the measurement of soil moisture suction. It has been developed over a number of years and is very similar in operation to the thermistor psychrometer which it effectively replaces. Improvements in performance have been made and the latest instrument can measure a much wider range of soil suctions in a much shorter time. Much of this improvement is due to extensive work on calibration procedures which have determined many of the characteristics of the instrument. It is also due to advances in micro-chip technology which allow amplification within the probe of the signal generated by the temperature depression of a water drop attached to a transistor.

Perhaps the greatest improvement, and the one that sets this instrument apart from any others measuring soil suction, is that the analogue output can be read by a standard millivoltmeter or logged by any millivolt data logger. The latter allows the storage, reduction and manipulation of the data to be carried out by a PC. Plotting of the output can be achieved in real time, with a logger designed for the instrument and a colour dot matrix printer. These advances have enabled the psychrometer to take its place in the modern soil mechanics laboratory.

Introduction

Development of the transistor psychrometer has taken place in several stages over the past 10 years. It has now progressed to the stage where it is being manufactured under an agreement with CSIRO in Australia and is being used in a number of laboratories around the world.

----

1 Principal, Woodburn Associates, Consulting Geotechnical Engineers, Adelaide, South Australia
2 Principal Research Engineer, VIC ROADS, Melbourne, Victoria, Australia
3 Principal Experimental Scientist, CSIRO, Division of Soils, Adelaide, South Australia

The initial concept and working models of transistor probes were made by Paul Peter of CSIRO in the mid-1980s. This followed earlier work using thermistors as temperature sensors to measure the temperature depression of the electronic equivalent of a wet bulb thermometer. When that research ceased, the manufacturing rights were granted to John Woodburn of Soil Mechanics Instrumentation (SMI) and that organisation carried out further work and development of the probe to the stage where prototype instruments were produced. Dr. Jim Holden and his research group at Vic Roads (an Australian State Road Authority) then used one of the prototype instruments to develop standard operating procedures which gave a high degree of reproducibility in the results.

Work on the new type of psychrometer began when it became apparent that with microchip technology, a probe using transistors as heat sensors could be used for measuring relative humidity. As now developed the probe of the psychrometer consists of a "wet "and "dry" transistor with the temperature depression of the wet transistor measured and amplified within the probe. The analogue output for most soil testing has been adjusted to lie within the range 6 to 500 millivolts representing soil suctions between pF 3.0 and about pF 5.0. For conventional use the instrument is calibrated to pF 5.0 but may be calibrated to higher suctions using special procedures.

The accuracy of the instrument is dependent on the degree of ambient temperature control during the period of the test. For this reason the probes are inserted into a thermally insulated bath. This ensures that the probes and specimens remain at a near constant temperature during the period of any one test. As the probes are affected by room air temperature changes, greater accuracy and reproducibility of results is obtained within a room controlled to about $\pm 0.5^{\circ}$ C. The degree of accuracy and reproducibility for soil suctions above pF 3.5 with this degree of temperature control and special operating procedures is close to $\pm$ 0.02 pF. Using standard operating procedures an accuracy of $\pm$ 0.05 pF is easily obtained.

When testing soil samples the time allowed for a test is the same as that allowed for calibration of the probes. At the present time this has been fixed at one hour. However indications are that this time for testing can be reduced without loss of accuracy. For most engineering purposes, an accuracy of $\pm$ 0.05 pF is all that is required. This order of accuracy allows room air temperature changes as high as $\pm$ 1.5°C to occur during the period of the test when testing soils above pF 4.0.

Current Techniques for Soil Suction Measurement

A number of techniques are available for soil suction measurement in the laboratory and these have been recently reviewed (Nelson and Miller, 1992). There are problems with many of the techniques due to lack of range, length of time required for equilibrium and their ability to measure only matrix suction.

Soil suction measurement is now recognised as one of the series of tests available to the geotechnical engineer practising in the field of expansive soils and Standards are available in at least two countries. The techniques currently in use have been available for many years (between 20 and 40) and have not really advanced in that time. This contrasts with present day laboratory procedures used for triaxial, swelling pressure and oedometer testing which use computer driven equipment for accurate monitoring of the test, the reduction of the results and the storage of data.

Psychrometric Technique

This technique is widely used in Australia where there is a Standard (AS 1289.2.2.1-1992) governing the test procedure. The most commonly used instrument is the Wescor Dew Point Microvoltmeter which is used with the C-51 sample chamber. The equipment is expensive in Australia, prone to damage in inexperienced hands and has an upper limit of about pF 4.5. This is often too low for testing soils sampled at the end of a drying cycle in semi-arid and arid areas. The sample chamber can suffer from problems with contamination of the thermocouple and subsequent corrosion unless care and periodic cleaning is carried out.

Filter Paper Technique

This technique has been adopted widely in the past few years and an American Standard (ASTM, 1990) is currently at the review stage. The technique has been available for many years but has recently undergone a revival with the advent of cheaper, more accurate balances and extensive work by a number of researchers. It is doubtful that the cost of a test is no more than that of a moisture content test as reported by McKeen (1992) because in addition to the moisture content determination involving measurements to 0.0001gm., it is a time consuming test. Monitoring of the filter paper weight over a period of at least a week may be required to ensure equilibrium has been reached and great care must be taken to ensure no moisture loss occurs prior to weighing. The technique is useful in that it has the ability to measure both matrix and total suction although with the latter a stable temperature environment must be provided as moisture transfer has to occur in the vapour phase. Consequently even longer times may be required for equilibrium to occur and great care must be taken to ensure that condensation inside the chamber does not affect the filter paper.

Development of the Transistor Psychrometer

Suction/Relative Humidity Relationships

A psychrometer is an instrument which measures relative vapour pressure or relative humidity. Relative humidity is related to soil moisture suction in accordance with the relationship:

$$s = Log[(1.284657 \times 10^6 + 4.703\ t) \times Ln(H/100)]$$

where s = suction in pF
                     t = temperature in °C
                     H = relative humidity specified as a percentage

This equation provides an indication of the change in suction with temperature and relative humidity (Table 1). When plotted for 20°C it gives the well known curve shown in Figure 1(a). As the most relevant part of this curve to engineers and scientists interested in the flow of moisture in soils lies above 95% relative humidity, an enlargement of the upper part is shown -Figure 1(b). What this enlargement shows, and what is often not appreciated is that any instrument using psychrometric techniques must be able to measure relative humidities up to 99.9%. It must also be able to differentiate between relative humidity changes of about 0.02% at pF 3.0 (ie a suction change of 0.1 pF). This represents a temperature depression of 0.002 °C for the wetted sensor in a pair used in a similar way to a wet and dry bulb thermometer.

SUCTION vs RELATIVE HUMIDITY AND TEMPERATURE

SUCTION DISPLAYED IN pF

RELATIVE HUMIDITY (%)	TEMPERATURE (DEGREES CENTIGRADE)					
	15	20	25	30	35	40
99.90	3.132	3.139	3.147	3.154	3.161	3.168
99.50	3.832	3.839	3.847	3.854	3.861	3.868
99.00	4.134	4.141	4.149	4.156	4.163	4.170
98.50	4.311	4.319	4.326	4.333	4.340	4.347
98.00	4.437	4.445	4.452	4.459	4.466	4.473
97.50	4.535	4.543	4.550	4.557	4.564	4.571
97.00	4.615	4.623	4.630	4.638	4.645	4.652
96.50	4.684	4.691	4.698	4.706	4.713	4.720
96.00	4.743	4.750	4.757	4.765	4.772	4.779
95.50	4.795	4.802	4.810	4.817	4.824	4.831
95.00	4.842	4.849	4.857	4.864	4.871	4.878
94.50	4.884	4.892	4.899	4.906	4.914	4.921
94.00	4.923	4.931	4.938	4.945	4.952	4.959
93.50	4.959	4.967	4.974	4.981	4.988	4.995
93.00	4.993	5.000	5.007	5.015	5.022	5.029
92.50	5.024	5.031	5.038	5.046	5.053	5.060
92.00	5.053	5.060	5.068	5.075	5.082	5.089
91.50	5.080	5.088	5.095	5.102	5.110	5.116
91.00	5.106	5.114	5.121	5.128	5.135	5.142
90.50	5.131	5.138	5.146	5.153	5.160	5.167
90.00	5.154	5.162	5.169	5.177	5.184	5.191
89.50	5.177	5.184	5.192	5.199	5.206	5.213
89.00	5.198	5.206	5.213	5.220	5.227	5.234
88.50	5.219	5.226	5.234	5.241	5.248	5.255
88.00	5.238	5.246	5.253	5.260	5.268	5.275

Table 1-Suction vs Relative Humidity and Temperature

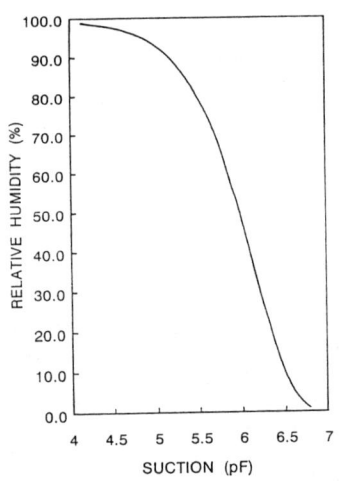

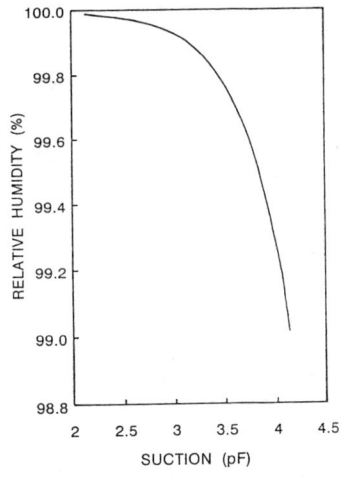

Figure 1-(a)Suction vs Relative Humidity at 20°C   (b) Enlarged

There are a number of electrical components which are temperature sensitive and which can be used to measure temperature changes of this order. Some of these components register the very small temperature changes in micro-volts (thermocouples and thermistors) while others can utilise a higher voltage source and register outputs in the millivolt range (transistors and the latest range of temperature sensors).

Transistors as the Sensing Device

The CSIRO has been actively engaged in the investigation of soil suction techniques for many years and early work on psychrometers was carried out in the 1960's using thermistors. With those sensors it was found that for reasonable accuracy and reproducibility of results the test had to be carried out in a room controlled to ± 0.5 °C with a constant temperature bath controlled to ± 0.001 °C (Richards, 1965). It was also found that there was a need to accurately control the water drop size on the "wet bulb" sensor.

The idea of using transistors instead of thermistors as the temperature sensors in the psychrometer probes was first conceived by the Soil Engineering Group of the CSIRO, Division of Soils in 1984. This was partly influenced by the fact that matched pairs of thermistors were expensive, each probe needed its own power supply (1.35V mercury cell), the glass sheathing enclosing the temperature sensitive element was fragile and therefore more easily damaged and they also deteriorated with time. On the other hand transistors and integrated circuits (operational-amplifiers) were relatively cheap, robust and reliable. Advances in technology had also improved the characteristics of op-amps at that time to the extent that it was both economical and feasible to use individual op-amps in each probe.

In the final development of the transistor psychrometer the two transistors (temperature sensors), the operational amplifier and other associated components were mounted on a printed circuit board designed to fit into the shaft of the probe (Figure 2). This ensured that with the probes placed in the thermally insulated bath, the components in the probe were buffered from temperature changes occurring in the laboratory environment (Figure 3). The transistors and other components that make up each probe are carefully chosen and matched to circuit requirements. Silicon NPN type transistors are used which have a nominal base-emitter temperature coefficient of 2 millivolts/°C. This signal is amplified approximately 1000 fold with the aid of a high input impedance operational amplifier, giving an output from the probe of 2 volts/°C. Output voltages of this order can be readily monitored with a number of devices ranging from digital voltmeters and pen recorders to data loggers.

Operating Procedures and Accuracy

Prototypes of the transistor psychrometers have now been in use in the Vic Roads and CSIRO laboratories for several years. During this time a number of techniques for calibrating the psychrometer and testing soil specimens have been examined. Specific procedures and conditions have now been established which provide a high order of accuracy (i.e. up to ± 0.02 pF) in suctions over about pF 3.5. However, it has been found that an adequate degree of accuracy for most engineering purposes can be achieved using less stringent procedures and conditions.

printed circuit board
probe body
'wet' transistor
sleeve
water drop
sampling tube
soil
sealing tape
'dry' transistor

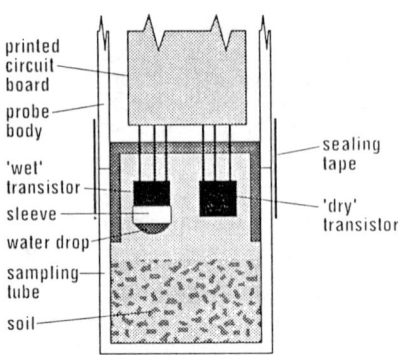

Figure 2-The Probe Tip Containing the Sensing Elements (Transistors)

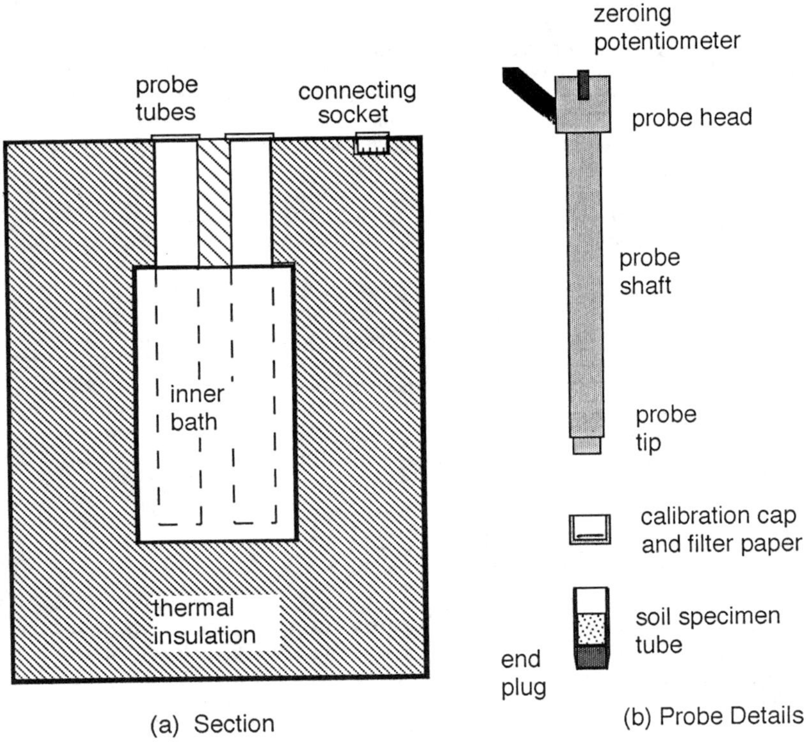

probe tubes
connecting socket
inner bath
thermal insulation
(a) Section
zeroing potentiometer
probe head
probe shaft
probe tip
calibration cap and filter paper
soil specimen tube
end plug
(b) Probe Details

Figure 3-The Probe and Thermally Insulated Container

Calibration

Calibration of the transistor psychrometer is required because the output from the instrument is in millivolts which must be converted to a suction value if the instrument is to be used for the suction testing of soil specimens. The procedure, described in detail by Dimos (1991), is similar to the calibration of other psychrometric instruments and involves the use of standard salt solutions prepared to give equivalent relative humidities between pF 3.0 and pF 5.0. Before calibrating, the probes are zeroed with a pF 2.0 solution. A typical calibration line for a probe is shown in Figure 4.

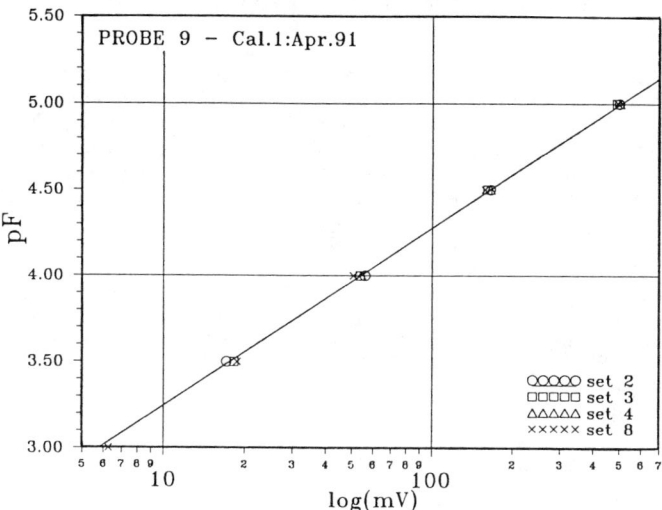

Figure 4-Calibration Line of Transistor Psychrometer Probe

The calibration line is affected to some extent by changes in the room or bath temperature, the drop shape and the size of the gap between the drop and soil surface, although these factors may only be critical when very high orders of accuracy are being achieved. Other factors are also thought to influence the calibration and it is for this reason that standard calibration procedures have been established.

Prior to calibrating and any other form of suction testing the probes must be zeroed or located on the calibration line using standard salt solutions. Early work on calibration techniques used distilled water for this purpose (H=100%) but after much work and finally dispensing with free liquids because of contamination problems, impregnated filter papers were accepted for both zeroing and calibration. When the testing program allows zeroing to be carried out overnight, filter papers impregnated with pF2.0 solution (H=99.99%) are used. Even this very high relative humidity gives a small positive output from the probe giving a more definite "zero" reading

Calibration of the probes is carried out using the PVC caps (Figure 3) which hold the filter paper discs impregnated with standard salt solutions. Before calibrating the probes, each tip must be cleaned with distilled water and inspected to ensure that the sleeve on the 'wet' transistor is at the specified height to hold the water drop (Figure 2).

Each calibration cap is prepared by placing a two-layer disc of filter paper, 15mm diameter in the base. The filter paper is then saturated using 3 drops of pF 2.0 salt solution and the cap sealed with a rubber stopper until all have been prepared. The caps are then placed on the end of each probe in turn and carefully inserted into the thermally insulated bath, after placing a standard drop of distilled water on the "wet" transistor. At this low suction the probes require many hours to reach equilibrium and are generally allowed to run overnight. In the morning the probes are zeroed prior to beginning the calibration which follows the same procedure outlined above, but using pF 3.0, 3.5, 4.0, 4.5 and 5.0 solutions in turn. For each stage the data is collected for one hour. This has been found an adequate time for equilibrium to occur and to obtain a result (Dimos, 1991).

For each calibration test using a particular concentration of salt solution a graph of millivolts *versus* time is obtained. The equilibrium value obtained after one hour is the value used for the calibration line. At the completion of a set of tests using the range of standard solutions, a calibration line of pF *versus* log millivolts can be produced for each probe. This entire procedure can be conducted in a normal working day but should be repeated at least two times initially to ensure that the calibrations have been carried out correctly. When a number of sets of results have been obtained, an average calibration line can be determined for each probe.

Check calibrations should also be conducted at regular intervals once the psychrometer is in use with the maximum recommended period between them being 3 months. A calibration check should also be carried out immediately before soil testing if a very high degree of accuracy is required. A calibration line is supplied for each probe with the equipment and the calibrations obtained soon after receiving the equipment should be compared with these.

Preparation of soil specimens

Soil samples are usually obtained in sample tubes during auger drilling in the field. In the laboratory a hydraulic jack is used to push the sample out of the tube after removing the seals. When preparing the specimens care must be taken to minimise exposure of the soil to the atmosphere.

About 5mm is shaved off the end of the sample to ensure that the specimens tested are not contaminated or dried out. A 10mm high sample ring (with the same diameter as the sample tube) is placed on top of the sample tube and the sample jacked up until it protrudes about 1mm above the sample ring. At this point the ring is released and the sample jacked again until there is about 5mm showing below the ring. A sample slice including the ring is then cut off and trimmed to ring height.

At least two specimens should be obtained from each slice to allow for comparison and averaging of the results. Each specimen sampling tube (Figure 3) is pushed into the sample slice either by hand or using a small press. The position of the specimen in the specimen tube is then adjusted accurately using a specimen spacer. End plugs which have also been specially made for each of the probes are inserted in the bottom end of the tube to keep the specimen in position. Rubber stoppers are inserted in the other end of the specimen tube to eliminate any possible moisture loss. When all specimens are prepared they are immediately placed in the constant temperature room or laboratory and allowed to stand for at least half an hour to reach temperature equilibrium and to humidify the atmosphere within the specimen tube.

Testing of soil specimens

Before testing the specimens the probes must be stabilised overnight using pF 2.0 salt solution. They are then zeroed in the morning prior to placing the soil specimens. After removal of the probe and the pF 2.0 solution cap, the soil specimens are placed on the end of each probe in turn and the probes returned to the thermally insulated bath. Monitoring of the output then continues for at least one hour or until the graph of millivolts *vs* time reaches constant output. This gives a millivolt value for each probe. From this the soil suction in the specimen is obtained using the average calibration line for the probe.

If more than one set of specimens are to be tested in one day then *it is important that wet soil samples are tested before the dry samples to avoid a hysteresis effect.* When testing is complete the probes are dried, a new water drop added and stabilised overnight on pF 2.0 solution ready for testing the next day.

If results to ± 0.02 pF are required then further conditioning of the probes is required prior to soil testing by running the psychrometer with a solution of pF 3.0 for one hour. If the soil samples are likely to have a soil suction above pF 3.5, the probes should also be conditioned for a further hour with pF 3.5 solution. This conditioning is necessary in order to remove or measure the effects of drift prior to carrying out the test. This drift can be compensated at the reduction stage of the results or eliminated by comparing the millivolt value obtained during the conditioning stage with that obtained from the calibration line and re-setting the output accordingly.

Time for Equilibrium

When calibrating, the value after one hour is recorded as it is usual that a constant, or near constant reading has been achieved in that time. When testing soil specimens the reading is also taken after one hour although a longer time may sometimes be required for equilibration.

Temperature Control

The output from the probes varies with the temperature of the probes and these temperature changes are manifested as a drift in the output record. For this reason there must be a good degree of ambient temperature control during the period of the test. The purpose of the thermally insulated bath is to hold the body of the probes at a near constant temperature during the hour that the test is run. However because the head of the probe is outside the bath the probes can still be affected by room air temperature changes which produce temperature gradients along the probe shaft. It has been found that the greatest accuracy and reproducibility of results is achieved with the room temperature controlled to ±0.5°C.

The latest version of the transistor psychrometer is portable, contains 8 probes and has an insulated lid which can be closed after inserting the probes. This assists in keeping the heads of the probes at a constant temperature during the period of the test.

Accuracy

With good laboratory temperature control and the current operating procedures the transistor psychrometer is capable of measuring the total suction of a soil in the range of at least pF3.0 to pF5.0 with an accuracy of about ± 0.02 pF above pF3.5. This accuracy is greater than that required for most engineering applications. It is much greater than that specified by the Australian Standard AS 1289.2.2.1 - 1992 which states that results below pF 3.6 should be given to ± 0.1 pF and above pF 3.6 to ± 0.05 pF. To provide a check on the accuracy of the results and to enable an average to be calculated, specimens are usually tested in pairs or preferably in groups of three for greater accuracy.

Output

At least 48 specimens can be tested using the 12 probe psychrometer in a normal working day. This includes any initial conditioning of the probes with salt solutions at pF 3.0 and 3.5 and then performing 4 runs with soil samples. Without the conditioning procedures, even more samples can be tested although towards the end of the day the water drop may be depleted if most of the soils are at high suctions.

Test Monitoring, Reduction and Storage of Results

There are a number of means of obtaining and reducing the results produced by the transistor psychrometer and three options are shown schematically in Figure 5 and outlined below:

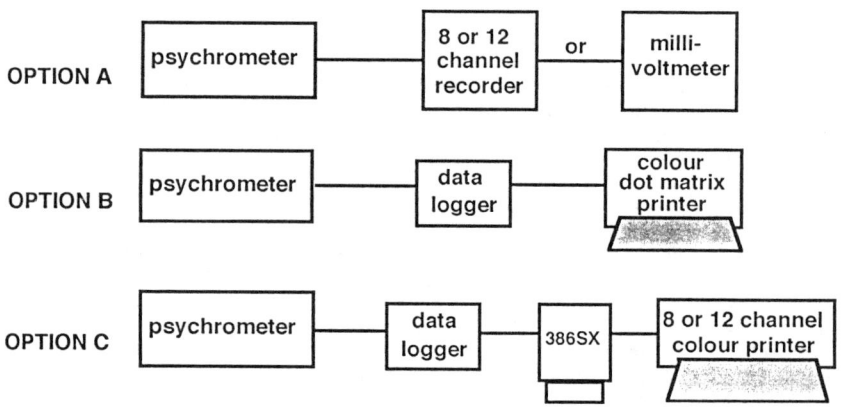

Figure 5-Psychrometer Data Recording and Manipulation.

Option 1. This system is the most simple. The results are obtained as a direct analogue output from a millivoltmeter or a millivolt recorder. The results obtained after 1 hour are recorded and the soil suction reduced from the set of calibration lines held in the laboratory.

Option 2. This is the system most commonly used. As supplied by SMI it consists of the psychrometer and data logger which has been programmed to use one of the new range of dot matrix colour printers. In the portable version the data logger is contained in the same thermally insulated case as the 8 probes and provided with printer and RS232 ports. With both systems the user has the option of a numerical or graphical output of the millivolt reading. The logger is programmed to show results from -23 to +1000mV, i.e. up to about pF 5.5. The results can be converted to suction values using the calibration lines obtained for each probe. They can also be accessed by a 386SL Notebook computer using appropriate software.

Option 3. This is the most sophisticated of the three options. It requires a software package which collects the data, presents the results on the screen during testing and allows the user to manipulate them. They can also be printed out in a graphical format with the important values given numerically once the test is complete. One such package has been written specifically for the Vic Roads psychrometer (VICroads 1992) and provides facilities to collect and store data, enable past data to be retrieved and to display the data collected in a number of formats. The procedure involved in the use of the computer package is described in detail by Dimos (1991).

Discussion

It has now been established that the transistor psychrometer is an accurate device for the measurement of the total water potential of soils. Work on the instrument is continuing with tests now being carried out on a more portable 8-channel model. It is believed that this will be suitable for field laboratory applications where high accuracy below about pF 3.75 is not required. One of the potential uses of this portable instrument lies in the control of the moisture condition of road fills and subgrades at the time of placement and compaction. These materials are often of high reactivity to moisture changes and should be placed at their equilibrium suction value. This would help to reduce the pavement cracking which often occurs with roads built on these soils.

Another possible use of a more portable instrument is in the determination of suction profiles without the need for actual sampling. Indications are that the technology now developed can be placed within a penetrometer and suction values obtained by staging the probe at various levels as it is forced into the ground. This would allow a suction profile to be measured to a depth of at least 6 meters in a relatively short time.

Conclusions

During the past three years much knowledge has been gained in the operation, use and calibration procedures required for the transistor psychrometer. It is now regarded as a reliable instrument for the measurement of soil suctions to better than the order of accuracy required for engineering applications and the current Standards. The instrument should be readily accepted in the modern soils laboratory because the process of measurement and reduction of the results can be carried out using data loggers and PCs with either currently available or customised software.
There is the potential to extend the use of the instrument to such applications as the placement and compaction of expansive clays at their equilibrium suction value and the determination of suction profiles without the need for actual sampling. Work is continuing on the evaluation of these applications.

Acknowledgements

The assistance of electronics engineer Bruce Lucas in the development of the SMI logger is greatly appreciated.

The second author J. Holden wishes to give thanks to his God the Father through the Lord Jesus Christ who he believes is the giver of all knowledge, and without whom nothing of value would have been achieved. He is also indebted to the considerable assistance of students Fiona Marmaras, Alexandra and Maria Dimos, Chin Phan and Josephine Whiting and many members of the staff of VIC ROADS. The help of David Veith and staff of the Vic Roads Electronics Laboratory and Michael Ma is gratefully acknowledged. This paper is presented with the Permission of Mr. Reg Patterson, Chief Executive VIC ROADS.

The views expressed are those of the authors and do not necessarily represent those of their parent organisations.

References

AS (1992), Soil Moisture Content Tests-Determination of the Total Suction of a Soil-Standard Method. AS 1289.2.2.1-1992, Standards Association of Australia, Sydney.

ASTM (1990), Proposed Standard Method of Test for Measurement of Soil Potential (Suction) Using Filter Paper.

Dimos, A., (1991). Measurement of Soil Suction using Transistor Psychrometer. VIC ROADS, Melbourne. Internal Report No. IR/91-3, September.

McKeen, R. G., (1992). A Model for Predicting Soil Behaviour. Proceedings of the 7th. International Conference on Expansive Soils, Dallas,USA

Nelson, J. D., and Miller, D. J., (1992). Expansive Soils. John Wiley and Sons, Inc., New York.

Richards, B. G., (1965). Measurement of the Free Energy of Soil Moisture by the Psychrometric Technique. Moisture Equilibria and Moisture Changes in Soils Beneath Covered Areas. A Symposium in Print. Butterworths, Australia.

Whiting, J., (1992). The Effects of Temperature on Psychrometer Performance. VIC ROADS, Melbourne. Internal Report No. IR/92-4, November.

MODELLING THE BEHAVIOUR OF COMPACTED SOILS

E.E. Alonso[1], A. Josa[1], A. Gens [1]

Abstract

Predictions based on a recently developed model for unsaturated soil behaviour have been compared with some results of two experimental programs on compacted soils recently published in the literature. In both programs compacted specimens were subjected to loading and wetting sequences. It is shown that experimental observations are consistently reproduced by the model. This success has led to a simplified framework for compacted soil behaviour in which model parameters represent soil type and compaction method whereas the as compacted dry density and moulding water content are conveniently described by initial preconsolidation stress and water suction.

Introduction

The concepts behind compaction were firmly etablished in the papers published by Proctor (1933) and have not experienced significant revisions. Two variables, dry density and compaction water content, are universally used to describe the compaction state of a given soil. A third component, microstructure, is also widely recognized as an important piece of information to explain the behaviour of compacted soil despite the fact that it lacks a simple quantitative descriptor to be used in practice. Field experience and an extensive laboratory research carried out in many parts of the world indicates that compacted soils may either compress (collapse) or swell when saturation increases. The first type of behaviour is inconsistent with the tenets of a single effective stress. On the other hand swelling upon wetting is strongly dependent on the particular sequence of loading and wetting imposed to the specimen. These conceptual difficulties explain the widespread use of empirical approaches and descriptive case-oriented studies when the behaviour of compacted soils is analyzed.

Two important concepts to understand the behaviour of compacted soils are the negative pressure (or suction in more general terms) of the pore water and the relationship between water and microstructure. The significance of these concepts will be described in the next section. They are a key part in recent efforts of the authors to provide a consistent framework to describe the behaviour of partially saturated soils (Alonso, Gens and Hight, 1987; Alonso, Gens and Josa, 1990; Josa, Balmaceda, Gens and Alonso, 1992; Gens and Alonso, 1992). Compacted soils belong to this class of soils. In fact, most of the experimental basic research on unsaturated soils has been carried

---

[1] Professor of Geotechnical Engineering. Civil Engineering School. Gran Capitán, s/n – Edificio D-2. 08034 Barcelona. Spain

out on statically compacted specimens (and less frequently on samples subjected to repeated impacts or kneading action). The mentioned frameworks have evolved towards constitutive models which not only explain the qualitative features of unsaturated soils behaviour but are capable of reasonable quantitative predictions.

The vast accumulation of experimental results concerning compacted soils offers a particular challenge to the developed models, if they are to be used as a predictive tool. In this regard the models should be able to reproduce main features of behaviour and the relationship between model parameters and compaction variables should be explored. This is essentially the purpose of this paper.

Review of the 'LC' constitutive model for unsaturated soils

Under isotropic stress conditions the main features of the model may be described in a (p, s) two dimensional stress space where $p = \sigma_m - u_a$ is the excess of total mean stress over air pressure and $s = u_a - u_w$ is the soil suction. When two specimens of a soil compacted at the same void ratio but different water contents (dry or wet of optimum) are loaded in compression, the dry-of-optimum specimen exhibits a larger apparent preconsolidation stress and a stiffer response than the wet of optimum counterpart. In terms of the adopted stress variables, the drier specimen has a higher suction that the wetter one. As a consequence, one may conclude that suction contributes to increase the apparent preconsolidation stress. In a (p, s) stress plane, the locus of apparent preconsolidation stresses induced by suction in samples compacted at a common dry density will plot in a curve such as LC_1 in Figure 1a. Within the framework of hardening plasticity, LC may be viewed as a yield surface. If the soil is saturated suction is zero and the corresponding preconsolidation stress has been named p_{o1}^*. The stress paths of two samples loaded in compression at two different suctions s_A and s_B are indicated in Figure 1a. Figure 1b shows the expected behaviour of the two samples in terms of changes in void ratio. Note that loading beyond the current preconsolidation stress will induce yielding of the soil and the accumulation of irreversible (plastic) volumetric strains. This yielding will move the current preconsolidation locus (LC_1) to a new position (LC_2) characterized by a new saturated preconsolidation stress p_{o2}^*. The LC yield surface bounds an "elastic" zone. Inside this zone wetting induces volumetric expansion as a result of the stress release associated to a decrease in suction. Similarly if suction increases (drying) the volume is reduced.

This model predicts also a collapse of the soil structure when a wetting path such as $B \rightarrow C_1 \rightarrow C_2$ crosses the current LC curve. Along this path the yield locus will be dragged towards position LC_2 which is characterized by a larger saturated preconsolidation stress p_{o2}^*. This increase in preconsolidation stress $(p_{o1}^* \rightarrow p_{o2}^*)$ is the result of a volumetric compaction of the sample (collapse). The LC yield locus explains therefore two basic features: the loading behaviour at different suctions and the collapse phenomena. This is the origin of the acronym LC (Loading-Collapse). The preceding ideas imply that, in relative terms, a given unsaturated soil will tend to expand when wetted at low confining stresses and to collapse at higher confining loads. This is for instance the case of the clayey sand compacted dry optimum, tested by Lawton, Fragaszy and Hardcastle (1989, 1991) (see Figures 4 and 5). Some results of this testing program will be compared with model predictions later in the paper.

Note that along the wetting path BC_1C_2 the model predicts a first stage $(B \rightarrow C_1)$ where swelling occurs. This tendency is, in the case depicted in Figure 1c, later reversed in $C_1 \rightarrow C_2$ until a net collapse is finally reached. In soaking tests, only the final volumetric deformation of the sample is often recorded and there is no indication of the effect of suction changes in the volumetric behaviour of the soil. However, a number of carefully conducted suction controlled oedometer or triaxial tests on collapsible compacted samples of different soil types have been reported in which a (small) amount of swelling is measured during the first stages of suction reduction. Authors reporting this behaviour are: Escario and Saez (1973), who tested samples of clay compacted dry of

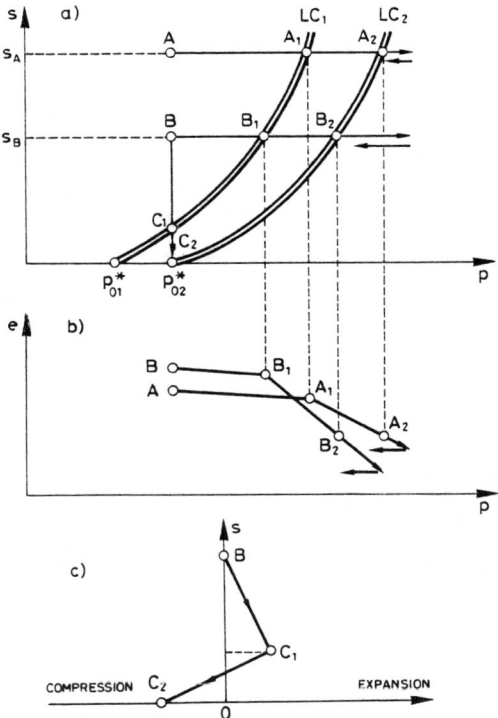

Fig. 1. Conceptual model. a) Yield curves and stress paths. b) Compression curves.
c) Deformation in a wetting path

optimum at 90% relative Proctor density; Cox (1978), who performed oedometer tests
on samples of a compacted marl taken from a highway fill at widely different dry densities; and Wheeler and Sivakumar (1993) who tested samples of statically compacted
kaolin dry of optimum at very low dry density ($\gamma_d = 1.20 Mg/m^3$; $e_o = 1.21$). The
change in specific volume along time of several kaolin samples brought to zero suction
under an isotropic stress of $40 kPa$ is shown in Figure 2. Note that a first swelling stage
is followed by a collapse of the specimens.

An interesting field case in this regard was provided by Charles, Hughes and Burford
(1984) who measured the settlements and internal deformation of a weakly compacted
fill made of sandstone and siltstone fragments. The fill was $70m$ deep and was placed
to restore an opencast mine. It experienced later a slow water table rise. Extensometer
data published by Charles et al. (1984) was reinterpreted (Alonso, 1993). Measured
strains were plotted in terms of the distance between extensometer locations and the
water table (which is a measure of the capillary height or suction at these points). The
data showed that for low confining pressures a small heave was measured during water
table elevation. At higher stresses important collapse ensued.

An important consequence of the model is that, given an initial structure of the
soil, a more compressed state may be reached either in a loading or in a wetting path.

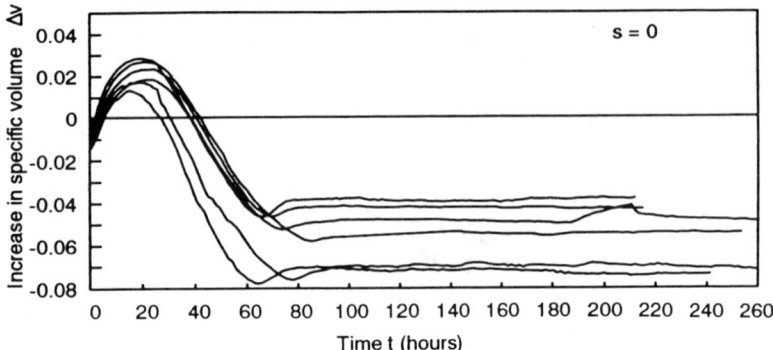

Fig. 2. Change of specific volume during equalization stage when several compacted kaolin samples are brought to zero suction under an isotropic stress of $40kPa$ (Wheeler and Sivakumar, 1993)

This is schematically indicated in Figure 3. The left part of the figure indicates a macrostructure of a soil in which granular sand-like particles and aggregates of clay platelets are in equilibrium with water under suction s and an isotropic stress state given by the total net mean stress p_1. Water partially fills the connected pore space. However, clay aggregates are almost certainly saturated . Total water suction inside these aggregates must be in equilibrium with the capillary suction which prevails in the larger pores, and helps to maintain the rigidity of these aggregates.

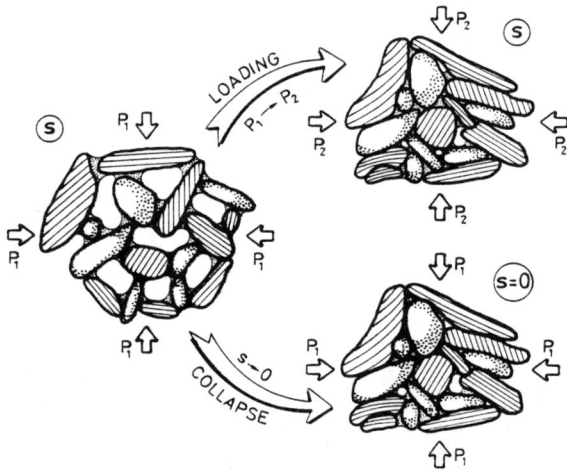

Fig. 3. Microstructural interpretation of loading and collapse deformations

If suction is reduced to zero (lower path of the figure) two mechanisms contribute to volume change reduction: a rearrangement of the structure and the distorsion of the aggregates which are unable to resist the shear induced by contact forces, once the suction is reduced. Some bending of clay particles has also been sketched as a possible mechanism for partial elastic recovery if the external load p_1 is eventually removed. Loading the specimen at constant suction (upper path in Figure 3) leads essentially to the same mechanisms of deformation. The maintained suction is now unable to resist the increased values of concentrated forces between particles. Rearrangements and distorsion of aggregates will again result. The LC yield locus represents the limit arrangement of particles and shape of aggregates which is in equilibrium with a given combination of mean stress, p, and suction, s. Any increase in p or reduction in s beyond the frontier given by LC provokes an irreversible rearrangement and distorsion of particles and hence marks the begining of a collapse. A detailed exploration of the consequences of the outlined framework has been given elsewhere (Alonso et al., 1987, 1990).

The following parameters are part of the model outlined before:

- Parameters associated with the LC curve: $\lambda(0)$, compressibility coefficient for the virgin loading curve; κ, elastic stiffness parameter against changes in total stress; κ_s, elastic stiffness parameter against changes in suction; r, ratio of minimum value of compressibility coefficient (for high values of suction) to $\lambda(0)$; β, controls the rate of increase of stiffness with suction, $(\lambda(s) = \lambda(0)[(1 - r)exp(-\beta s) + r])$.

- Parameters associated with changes in shear stress and the shear strength: G, shear modulus in the elastic domain; M, slope of the critical state line (related to the angle of internal friction); k, parameter which controls the rate of increase of apparent cohesion, p_s, with suction $(p_s = ks)$.

Compacted soil behaviour and model predictions

Lawton et al. (1989) presented data on collapse, under oedometric conditions, of a compacted clayey sand. In a second paper, Lawton et al. (1991), the effect of stress ratio on collapse was examined by means of a double triaxial test procedure. All of these tests were run without suction control. The second experimental program was carried out by Balmaceda (1991) and involved conventional oedometer collapse tests on a dynamically compacted silty sand. Some results of these tests will be compared with model predictions.

The aim of this comparison is twofold: firstly, it will illustrate the capabilities of the model outlined above to make quantitative predictions and to capture the essential features observed in the experimental results. Secondly, an effort will be made to relate conventional compaction variables to model parameters directly linked to the constitutive behaviour. This will allow a discussion on general trends of behaviour of compacted soils.

Tests performed by Lawton et al. (1989, 1991)

In order to model the experiments some parameters had to be established using as 'data' the same experimental results which are to be predicted. In addition, the fact that the loading and wetting steps of the experiments were run without suction control introduces difficulties because suction variation is not known. In fact, during the loading part of the experiments suction changes as a result of the condition of constant water content. As a consequence, suction decreases continuously during deformation, and a combination of collapse associated with suction reduction and loading-induced deformations coexist along the loading path. An additional state function, relating suction and water content, is therefore required to solve, in a coupled way, this type of condition which is typical in conventional oedometer testing.

Following previous developments (Lloret and Alonso, 1985; Alonso et al., 1990), the degree of saturation S_r was related to suction through the equation

$$S_r = 1 - m \, tanh(ns) \qquad (1)$$

where m and n are constants. In all the calculations performed the following parameters have been selected: $m = 1$; $n = 1 MPa^{-1}$. These parameters imply that saturation reduces to zero if suction increases indefinitely and that a suction of 1 MPa brings the degree of saturation to 0.24.

Lawton et al. (1989) tested compacted samples of a slightly expansive clayey sand ($w_L = 34\%$; $I_P = 15\%$; $Activity = 1$) at various dry densities and moulding water contents. Relative density was expressed as a percent of relative compaction with respect to the maximum Modified Proctor density. The (γ_d, w) optimum pairs were as follows: $(\gamma_d, w)_{Std.Proctor} = (1.81g/cm^3, 15\%)$; $(\gamma_d, w)_{Mod.Proctor} = (2.02g/cm^3, 10\%)$. They correspond to an 80% degree of saturation.

Virgin consolidation stiffness ($\lambda(0)$) was approximated on the basis of the Plasticity Index of the soil using the empirical relationship (Atkinson and Bransby, 1978): $\lambda(0) = \frac{I_P\gamma_s}{2\gamma_w ln10}$. For $I_P = 15\%$ a value $\lambda(0) \simeq 0.09$ is found for $\frac{\gamma_s}{\gamma_w} = 2.73$. Within the elastic range a value of $\kappa = \lambda(0)/10$ was adopted. Once $\lambda(0)$ and κ are established, the remaining parameters controlling the LC yield surface (r, β and p^c) are particularly relevant to adjust model predictions. They were selected, after a trial and error procedure, in order to reproduce the collapse vs. overburden pressure relationship for a sample compacted at 80% relative compaction and 13% moulding water content (Figures 3 and 4 of Lawton et al., 1989, paper; see also Figure 4). It was adopted $r = 0.4$; $\beta = 0.8 MPa^{-1}$; $p^c = 5kPa$. The value of r reflects a moderate collapse potential of the soil and the value of β indicates a slow increase in stiffness with suction. For the reference stress, p^c, a small value was selected.

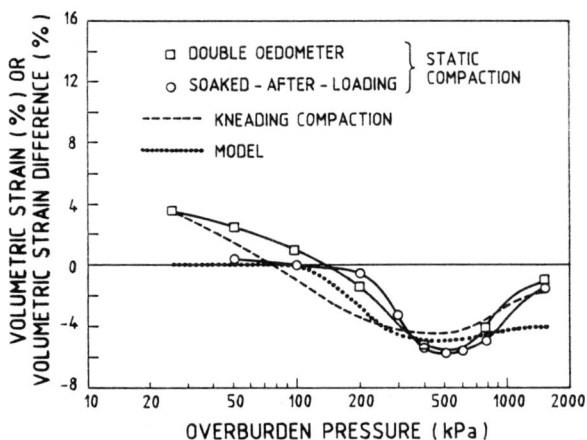

Fig. 4.- Comparison between model calculations and oedometer test results reported by Lawton et al. (1989). (Relative compaction: 80% Moulding water content: 13%)

Swelling behaviour is controlled by κ_s. A small value, $\kappa_s = 0.001$, characterizes a slightly swelling soil. The behaviour in shear should also be specified since loading and collapse under oedometric conditions involves variations in shear stresses. The following parameters, characterizing shear behaviour, were adopted: $G = 5MPa$; $k = 0.8$ and $M = 1.2$. The value of M (slope of the Critical State Line) was calculated for $\varphi' = 30^o$, a value mentioned in Lawton et al. (1991) for the soils being tested.

Finally, the initial state of the sample as described by the preconsolidation stress for saturated conditions, p_o^*, should be specified. Since the increase in preconsolidation stress with suction (as given by parameter β) is relatively small, the overburden pressure for which collapse begins to manifest gives an indication of the value of p_o^*. In fact p_o^* should be somewhat smaller than the overburden pressure which marks the beginning of collapse. Based on the results plotted in Figure 4 a value $p_o^* = 100 kPa$ was adopted. The values of initial void ratio and degree of saturation could be derived from the data given in the papers.

Model predictions for the set of parameters indicated have been plotted in Figure 4. The paper by Lawton et al. (1989) reports also the effect of moulding water content on volumetric strain differences measured through double oedometer techniques. Their results, reproduced in Figure 5, correspond to an 80% relative compaction. Samples were, in this case, compacted by repeated impact. Model predictions for the same set of parameters have been plotted in Figure 5.

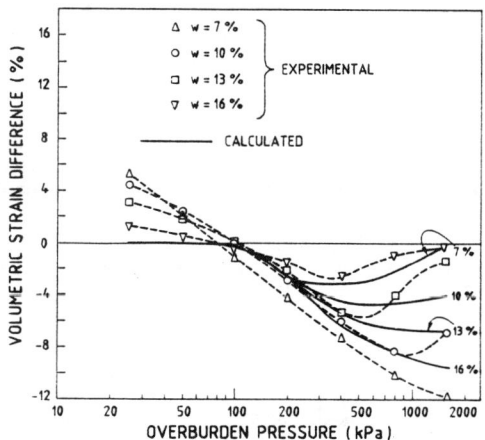

Fig. 5.- Comparison between model calculations and double oedometer tests for samples compacted at 80% relative compaction and different moulding water contents. (Test results taken from Lawton et al., 1989)

Several aspects may be discussed with regard to Figures 4 and 5:

a) Experimental results in Figure 4 show that the soaking-after-loading technique in the swelling range, leads to considerably lower deformations than the double oedometer technique. This is a manifestation of the stress-path dependency of swelling behaviour in expansive soils which has been pointed out by some authors (Justo, Delgado and Ruiz, 1984; Sridharan, Rao and Sivapullaiah, 1986). The model presented before has no capabilities to reproduce this stress-path dependency. Although a new framework for soils which exhibit a distinctive swelling behaviour has been formulated in Gens and Alonso (1992).

b) The model reproduces the trends observed in the collapse part of the experiments. In particular, it predicts in a natural way that the maximum of collapse, for decreasing initial water contents, is measured at increasingly larger loads. In fact, the maximum of collapse, according to the theoretical framework, should correspond for a given suction (or water content) to the preconsolidation stress. This stress is very much related to the compactive effort specially if the increase in stiffness with suction is moderate. This comment is also supported by the experimental findings in Lawton et al. (1989). Note also that even though the model predicts a continuous increase in collapse with applied stress, a maximum of collapse has been computed in all the tests reproduced (Figure 5). This is explained by the fact that loading under constant water content reduces continuously the water suction. The results in Figure 4 and 5 show, however, that the model is, in general, unable to reproduce the marked final reduction in collapse deformation which follows the maximum. This indicates that the soil tested exhibits probably a fundamental maximum of collapse within the range of applied overburden stresses. A modification of the LC model to account for this type of behaviour has been proposed in Josa et al. (1992).

An interesting conclusion of the effect of stress ratio on collapse (Lawton et al., 1991) is that the volumetric strain depends on the applied mean stress but not on the stress ratio. This question may be also analyzed with the help of the model. It can be shown that the maximum volumetric strain difference for a specific initial (p_i, s_i) is given by the following expression:

$$\Delta \epsilon_{v\ max} = \frac{\lambda(0) - \kappa}{v}\ ln\left[\frac{1}{2}(\frac{\xi + 2}{\xi + 1})^{\frac{\lambda(s)-\kappa}{\lambda(0)-\kappa}}\right] \tag{2}$$

where $\xi = p_s/p_i$. An extreme of equation (7) is found when $s \to \infty$ and $p_i \to 0$. In this case, if the soil parameters adopted are substituted in (7), it is found $\Delta \epsilon_{v\ max} \simeq 3\%$ which indicates a minor effect of stress ratio on volumetric strain.

Maintaining unchanged the model parameters a final check has involved the comparison between predicted axial and radial strains induced by collapse and the measured values, always through the double triaxial technique. The results have been plotted in Figures 6 and 7. Despite some discrepancies, the model seems to capture appropriately the overall behaviour.

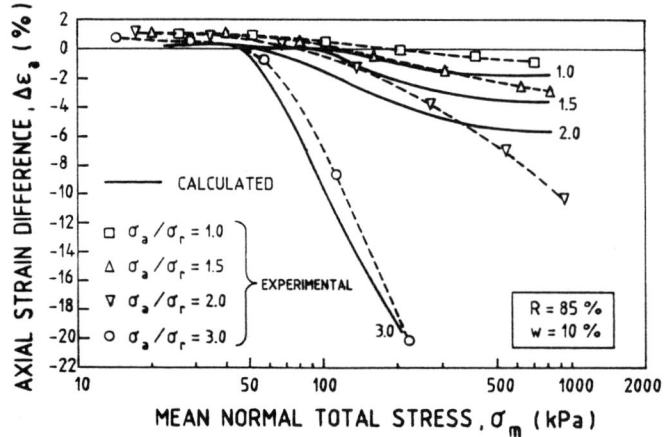

Fig. 6.- Comparison of calculated and measured axial strain differences as a function of mean normal total stress. (Test data after Lawton et al., 1991)

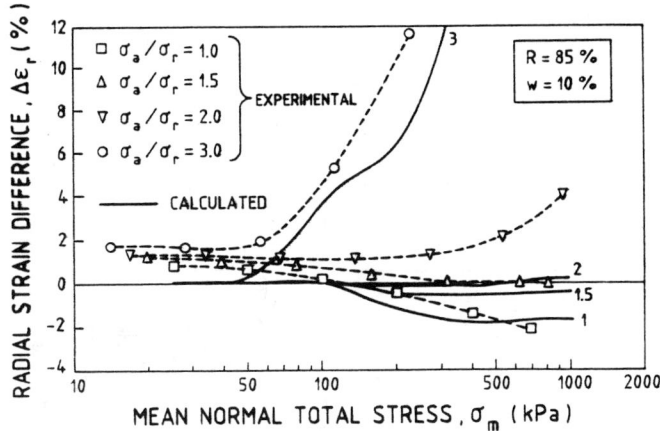

Fig. 7.- Comparison of calculated and measured radial strain difference as a function of mean normal total stress. (Test data after Lawton et al., 1991)

Tests performed by Balmaceda (1991)

Balmaceda (1991) tested samples of a nonplastic silty sand taken from the core of an earthdam under construction. Samples were dynamically compacted in a Standard Proctor mould. Seven hammer blows were applied to compact the samples. Several groups of identical samples were prepared for different initial conditions, as defined by (γ_d, w), in order to perform oedometric loading-wetting tests. During application of the load water content was kept constant. Once equilibrated, the samples were soaked and the volume change measured. The measured results for different groups of samples have been plotted in Figures 8a, b, c, d. These groups correspond to samples tested at increasing initial water content: 8.62%; 10.52%; 13.39% and 16.93%. The plots in Figure 8 show the initial loading at constant water content, the effect of soaking (a collapse in all the cases except for the case plotted in Figure 8d), the subsequent loading once the sample is saturated and the final unloading. Specimens were soaked at vertical loads ranging between $0.02MPa$ and $0.8MPa$. The maxium applied vertical stress was $0.8MPa$.

In this case model parameters were derived from the group of tests represented in Figure 8a. These parameters were maintained in the remaining tests analyzed. The slope of the saturated virgin compression line was measured to be $\lambda(0) = 0.07$. A small value for the unloading expansion index ($\kappa = 0.009$) was also found. Parameters r and β control the slope of the loading curve and therefore the amount of collapse. The values $r = 0.35$, $\beta = 1MPa^{-1}$ were selected. A small value $p^c = 0.6kPa$ was taken for the reference stress. The soil tested did not exhibit any measurable swelling so a very small elastic stiffness parameter for changes in suction was used: $\kappa_s = 0.001$.

This case, which involves, again, constant water content tests, requires the specification of a state function relating suction and degree of saturation. The relationship given by equation (5) was used with $m = 0.8$ and $n = 10MPa^{-1}$. This relationship could be found in this case because suction was measured in the compacted samples using a filter paper technique.

Concerning the shear part of the model the following parameters were selected:

$M = 1.2$ (corresponds to a friction angle $\varphi' = 30°$); $G = 0.1MPa$ and $\kappa = 0.8$ which represents a moderate increase in apparent cohesion with applied suction.

The saturated preconsolidation stress, p_o^*, can be inferred from the measured void ratio vs. vertical stress curves presented in Figure 8. This preconsolidation stress shows a continuous increase with the as compacted dry density.

A comparison between measured deformations and model predictions is given in Figure 8. The agreement is good in most cases.

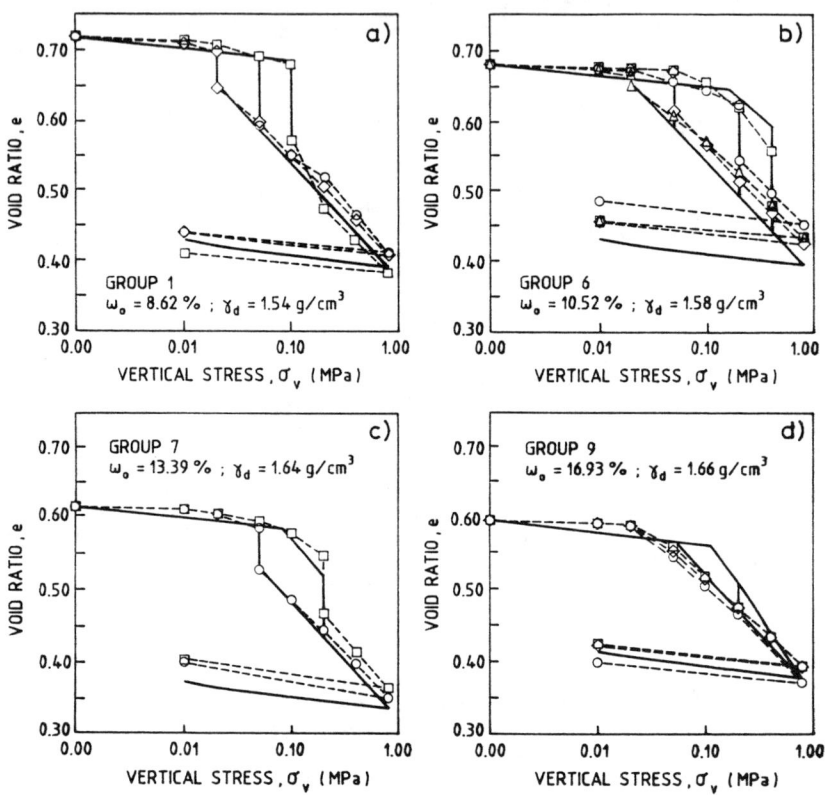

Fig. 8. Loading-collapse oedometer tests results and model predictions

Discussion

The results presented in the previous section show that the developed model offers an encouraging capability to reproduce the behaviour of compacted soils over a wide range of compaction conditions.

It has been shown that model parameters may remain constant for different compaction conditions and still predictions are reasonable. This comment is related to a key question which concerns the behaviour of compacted soils. An extreme position would suggest that for a given mineralogical content and grain size distribution, every combination of compaction method, compaction water content and density achieved would produce a unique soil which should be characterized, if a model of behaviour is available, by a set of constitutive parameters. This is equivalent to saying that specimens located at different points in the (γ_d, w) compaction plane would require different sets of constitutive parameters. An equally extreme but opposite view is to accept that a single set of constitutive parameters is able to represent different compaction conditions, the obvious differences among samples being attributed to different initial conditions. Within the context of the developed model this is equivalent to saying that variations in initial suction and saturated preconsolidation stress (p_o^*) would be enough to capture differences between compaction conditions.

The results presented previously provide some useful elements for the discussion. Differences in behaviour associated with different compaction conditions have traditionally been interpreted as a result of varying microstructures. This is a commonly suggested explanation for differences in behaviour when comparing dry or wet of optimum conditions. If changes in microstructure are relevant then a change in model parameters seems appropriate. However, it is very difficult to find the necessary evidence in many tests published, simply because the compacted specimens are tested by selecting, as initial conditions, the as compacted conditions. For instance, samples compacted wet of optimum are found to exhibit very low or negligible collapse properties unlike their drier counterparts compacted at the same dry density. But this difference in behaviour may be only due to the low water suction which accompanies wet of optimum compaction.

A precise check of the effect of structural differences requires the comparison of behaviour of samples compacted at the same initial void ratio and different water contents and tested under identical initial conditions and stress path. This type of test has not apparently been performed in the experimental programs analyzed previously. However it has been shown that for the type of soils considered, a single set of constitutive parameters leads to reasonable predictions wet or dry of optimum. One may conclude that for the soils tested by Lawton et al. (1989) and Balmaceda (1991) the microstructure induced by compacting dry or wet of optimum has a secondary role if compared with the attained void ratio. For these soils, as compacted conditions may be properly described by a pair of stress variables (p_o^* and s_o in the model described) which define the initial conditions. Otherwise a single set of constitutive parameters are capable of reproducing the loading and wetting behaviour of the soils tested.

Conclusions

A model which is able to reproduce many observed features of behaviour of moderately plastic to nonplastic compacted soils has been briefly described in the paper. The model is formulated within the framework of hardening plasticity and adopts two effective stress variables: excess of total stress over air pressure (net stress) and water suction. It has been shown that the model has good predicting capabilities to reproduce the behaviour of specimens in loading and wetting paths. The model offers also a conceptual explanation to some specific features of behaviour such as the effect of stress ratio on wetting induced collapse, which has been analyzed in some detail.

It has also been shown that successful comparisons between test results for samples compacted at varying water contents and model predictions may be achieved with a single set of constitutive parameters. Differences among samples are accounted for by a stress state which defines initial conditions: the preconsolidation stress for the compacted specimen once it is saturated and the as compacted suction. The former is controlled by the achieved dry density whereas the latter depends largely on initial degree of saturation.

The limits to this simplified framework remain to be established when experimental data on different type of soils (and specifically more plastic soils) could be analyzed.

References

Alonso, E.E. (1993). Effet des variations en teneur en eau dans les sols campactés. *Rev. Franç. Géotech.*, No. 62, pp. 7-21.

Alonso, E.E., Gens, A. and Hight, D. (1987). Special problem soils. General Report. *Proc. 9th European Conf. Soil Mech. Fdn Engng, Dublin*, 3, 1087-1146.

Alonso, E.E., Gens, A. and Josa, A. (1990). A constitutive model for partially saturated soils. *Géotechnique*, No. 40, 405-430.

Atkinson J.H. and Bransby, P.H. (1978). The Mechaniscs of Soils. An Introduction to Critical State Soil Mechanics. *McGraw-Hill.*

Balmaceda, A. (1991). Suelos compactados. Un estudio teórico y experimental. *PhD Thesis*, E.T.S. de Ingenieros de Caminos, Barcelona.

Charles, J.A., Hughes, D.B. and Burford, D. (1984). The effect of a rise of water table on the settlement of backfill at Horsley restored opencast coal mining site 1973-1983. *Proc. 3rd. Int. Conf. on Ground Movements and Structures*, pp. 423-442.

Cox, D.W. (1978). Volume change of compacted clay fill. *Clay Fills*, ICE, London, pp. 79-86.

Escario, V. and Saez, J. (1973). Measurement of the properties of swelling and collapsing soils under controlled suction. *Proc. 3rd Int. Conf. Expansive Soils*, Haifa, pp. 196-200.

Gens, A. and Alonso, E.E. (1992). A framework for the behaviour of unsaturated expansive clays. *Can. Geotech. J.*, 29, pp. 1013-1032.

Justo, J.L., Delgado, A. and Ruiz, J. (1984). The influence of the stress path in the collapse-swelling of soils at the laboratory. *Proc. 5th Int. Conf. Expansive Soils, Adelaide*, 67-71.

Josa, A., Balmaceda, A., Gens, A. and Alonso, E.E. (1992). An elastoplastic model for partially saturated soils exhibiting a maximum of collapse. Computational Plasticity. Owen, Oñate, Hinton eds., *Pineridge Press.*

Lawton, E.C., Fragaszy, R.J. and Hardcastle, J.H. (1989). Collapse of compacted clayey sand. *J. Geotech. Engng, Am. Soc. Civ. Engrs* 115, No. 9, 1252-1267.

Lawton, E.C., Fragaszy, R.J. and Hardcastle, J.H. (1991). Stress ratio effects on collapse of compacted clayey sand. *J. Geotech. Engng, Am. Soc. Civ. Engrs*, 117, No. 5, 714-730.

Lloret, A. and Alonso, E.E. (1985). State surfaces for partially saturated soils. *Proc. 11th Int. Conf. Soil Mech. Fdn Engng, San Francisco*, 2, 557-562.

Proctor, R.R. (1933). Fundamental principles of soil compaction. *Engineering News Record* 111: 245-248; 286-289; 348-351.

Sridharan, A., Rao, A.S. and Sivapullaiah, P.V. (1986). Swelling pressure of clays. *Geotech. Testing J.*, 9, 24-33.

Wheeler, S. and Sivakumar, (1993). Triaxial testing of unsaturated soils. *Unsaturated soils: recent developments and applications*, COMMET course, CEE program on continuing education, Barcelona.

ON THE ELASTO-PLASTIC BEHAVIOR
OF AN UNSATURATED SILT

Y.J. CUI and P. DELAGE

ABSTRACT

Due to technical difficulties related to the control and measurement of the suction in unsaturated soils, few experimental data on the mechanical behavior of those soils exist. The paper shows some results of a study made with a new osmotic controlled suction apparatus, in order to better identify the elasto-plastic behavior of an unsaturated silt. Therefore, special attention was given to volume changes and to compressive-dilatant behavior during shear, as a function of both suction and confining stress. An increase of both stiffness and strength with suction is shown, as well as the predominant effect of the soil compressibility, which is a function of the suction. A volumetric criterion is proposed in order to distinguish between elastic and plastic behavior.

INTRODUCTION

Compacted soil is the most widely used construction material in civil engineering, but its behavior is not fully understood, because of its unsaturated state. The suction which develops in unsaturated soils due to both capillary and water adsorption actions is technically difficult to measure and control, and relatively few experimental data on the mechanical behavior of those soils are available. Another difficulty is the measurement of the volume change of unsaturated soil samples, which cannot be managed in the same way as in saturated soils, because of the presence of air within the sample.

CERMES (Soil Mechanics Research Center), Ecole Nationale des Ponts et Chaussées, La Courtine, 93167 Noisy le Grand Cédex, France.

However, some recent state of the art reports based on the existing experimental data showed that, within the frame of the independent variables approach, some concepts of elasto-plasticity developed for soil behaviour could be extended, taking into account the effect of the suction (Alonso, Gens et al.1987).

This paper presents some results of an experimental study developed in order to better identify the elasto-plastic behavior of a compacted silt, with the help of an osmotic technique used for controlling the suction of the unsaturated samples in a triaxial apparatus; for that purpose, a special attention was given to volume changes during shear under controlled suction.

TRIAXIAL TESTING OF UNSATURATED SOILS

The first results of triaxial testing of unsaturated soils were provided by Bishop and Donald (1961), who tried to validate Bishop's extension of the effective stress concept to unsaturated soils. The axis translation method was adapted to a triaxial cell, with independent controls of both air (u_a) and water (u_w) pressures, which allowed the control of the suction $s = u_a - u_w$. Various authors used the same system in order to define shear strength of unsaturated soils (Gulhati and Satija 1981, Fredlund 1979...). A general criterion was proposed by Fredlund et al.(1978), defining a ϕ^b angle, according to the following equation:

$$\tau = c + (\sigma - u_a) \, tg \, \phi' + (u_a - u_w) \, tg \, \phi^b \qquad (1)$$

An interesting discussion concerned the possible variations of the friction angle with suction changes. An interesting comparison can be made with controlled suction shear box results (Escario and Saez 1986), which show an increase of the angle with suction. More recently, more attention was paid to the assessment of the validity of limit state and critical state concepts for unsaturated soils (Toll 1990 and Wheeler 1991). By using a double triaxial cell in order to monitor volume changes, Wheeler (1992) showed, on a compacted kaolinite, the validity of the critical state in the two following spaces: volume vs net stress $(\sigma - u_a)$ and suction $(u_a - u_w)$ space, and water content vs the same variables.

All the previously mentioned results have been obtained with the help of the axis translation method. However, an alternative method exists for controlling suction; it is based on an osmotic principle, and the use of a semi-permeable membrane and of a solution of organic molecules (Polyethylene glycol, PEG 20 000) which are big enough not to be able to cross the membrane. If a soil sample is put in contact with such a membrane behind which circulates the solution, it will be submitted to an osmotic suction which is a function of the concentration of the solution (Williams and Shaykewich 1969). Suctions values up to 1500 kPa may be reached with this technique. The first application to geotechnical engineering was made by Kassiff and Benshalom (1971) on an oedometer, in order to study the influence of the suction on the swelling of a soil. Komornik et al. (1980) used the technique on a hollow cylinder triaxial apparatus; Delage et al. (1987, 1992) adapted it on a classical triaxial sample, and also used an osmotic oedometer in order to study the effect of the suction on the compressibility of an unsaturated silt.

MATERIALS AND METHODS

Sample preparation

The soil studied is an eolian silt found to the east of Paris near the village of Jossigny. Its geotechnical properties are presented on table 1. The mineralogical composition of the clayey fraction determined from X-ray diffractometry showed the existence of illite, kaolinite and interstratified illite smectite minerals. However, the silt did not exhibit(ed) any swelling properties. The natural silt was first dried under ambient conditions, ground and passed in a 400 μm sieve, and carefully wetted to the standard Proctor optimum moisture content. The powder was then put in a 38 mm diameter cylindrical mould, and slowly compacted (rate 150 μm/mn) to the standard Proctor Maximum Dry density, with a double piston system. In an attempt to get a uniform density, the 76 mm high sample was compacted in three layers.

Table 1 : characteristics of the Jossigny silt

w_L (%)	w_P (%)	PI	%<2 μm	%>80μm	w_{OPM} (%)	γ_{OPM}(kN/m^3)	γ_s (kN/m^3)
37	1	18	34	4	18	16.7	27.2

The osmotic triaxial apparatus

The osmotic triaxial apparatus (Delage et al. 1987) is described on figure 1. In this apparatus, the sample is put in contact on both bottom and top with two semi-permeable cellulotic dialysis membranes; the cell base and piston are carved with concentric grooves through which the solution is circulated with the membrane resting on a fine sieve mesh. The piston is connected to the basis of the cell by two flexible tubings. The solution is contained in a closed circuit, composed of the serial connection of the base of the cell, the piston, a reservoir and the pump. The reservoir is chosen big enough in order to ensure an effectively constant concentration in spite of the exchanges occurring through the membranes between the sample and the solution. The bottle is closed by a rubber cap, pierced by three glass tubes; two of them allow the circulation of the solution, the third one is a capillary graduated tube for monitorig water exchanges between the solution and the sample. Its stabilisation before any mechanical operation indicates that the desired suction has been reached in the whole sample. An advantage of the osmotic technique for triaxial testing as compared to the axis translation method is the reduction of the drainage length to half the height of the sample, which is better for ensuring uniform suction conditions during shear.

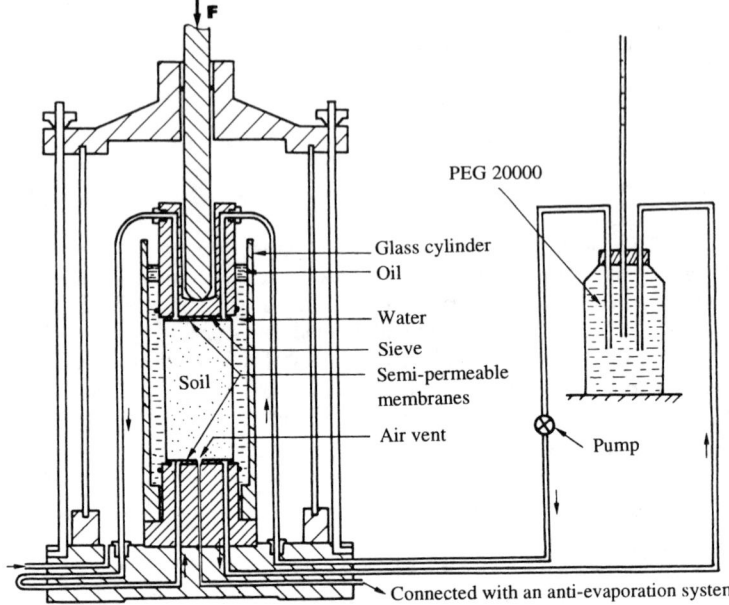

Figure 1 : The osmotic triaxial apparatus (Delage et al. 1987)

The volume change of the sample is monitored by a system similar to Bishop and Donald's system (1961), but with no mercury, for safety reasons (Cui 1993). A cylindrical glassy tube is placed around the sample, and filled with slightly coloured water; a thin layer of silicon oil is put on the water; the confining pressure is applied by air. The silicon oil avoids any dissolution of air in the water, and provides a good optical definition of the oil-air interface. Volume changes are optically monitored by following the changes of the level the oil-air interface with a cathetometer. A correction had to be made in order to take into account the penetration of the piston in the liquid and a special design of the piston was developed, as shown on figure 1.

Testing procedure

In order to accelerate the procedure, the initial suction imposition was not performed within the triaxial cell, but unconfined (Delage and Suraj de Silva 1992): the cylindrical triaxial sample was inserted in a semi-permeable tubing, and immersed in an agitated PEG solution of the desired concentration. In this case, the drainage length is equal to the radius of the sample, say 19 mm, and water exchanges are quicker than in the triaxial cell, were the drainage length is 38 mm. Suraj de Silva (1987) showed by monitoring of the weight of the sample, that equilibrium was reached after about one week .

After having reached the desired suction, the sample was placed in the triaxial cell, and the confining stress was applied. The experimental program is detailed in table 2; 14 samples were tested at 5 different suction values (0, 200, 400, 800, 1500), and at 5 different confining stresses (50, 100, 200, 400, and 600 kPa).

Table 2 : Experimental program

σ_3 (kPa)	s (kPa)	W_s (g)	e	S_r
50	0	143.4	0.694	98%
	200	144.7	0.621	77%
	400	145.4	0.623	74%
	800	145.3	0.619	68%
	1500	144.4	0.620	62%
100	200	145.4	*	*
	1500	144.2	0.612	64%
200	200	143.4	0.599	77%
	1500	143.3	0.621	64%
400	200	144.8	0.600	80%
	400	144.0	0.570	82%
	800	144.6	0.607	67%
	1500	143.8	0.619	66%
600	200	145.5	0.569	84%

(σ_3 is the confining stress, s the suction, W_s the dry weight, e the void ratio and S_r the saturation degree; no volume measurement for the * test) .

The shear rate was determined (Delage et al. 1987) in accordance with Ho and Fredlund's method (1982), and a similar value to that used for drained tests was adopted (1μm/mn).

Figure 2 shows the results (deviatoric stress and volume changes vs axial strain) of 5 shear tests made under a constant suction of 200 kPa and under confining stresses of 50, 100, 200, 400 and 600 kPa (no volume measurement for σ_3 = 100 kPa). Classically, both shear strength and compressive tendency increase with confining stress. For σ_3 = 50 and 100 kPa, the critical state seems to be reached rapidly, whereas some hardening is observed for $\sigma_3 > 400$ kPa, corresponding to a more pronounced compressive tendency.

For a higher suction (s=1500 kPa, figure 3), the general appearance of the various curves is quite different, which confirms the important effect of suction on shear behavior. The stiffening of the soil due to suction is observed when considering the initial slopes of the curves. For a same σ_3, shear strength is higher with a higher suction. Another important point concerns the existence of a maximum shear strength

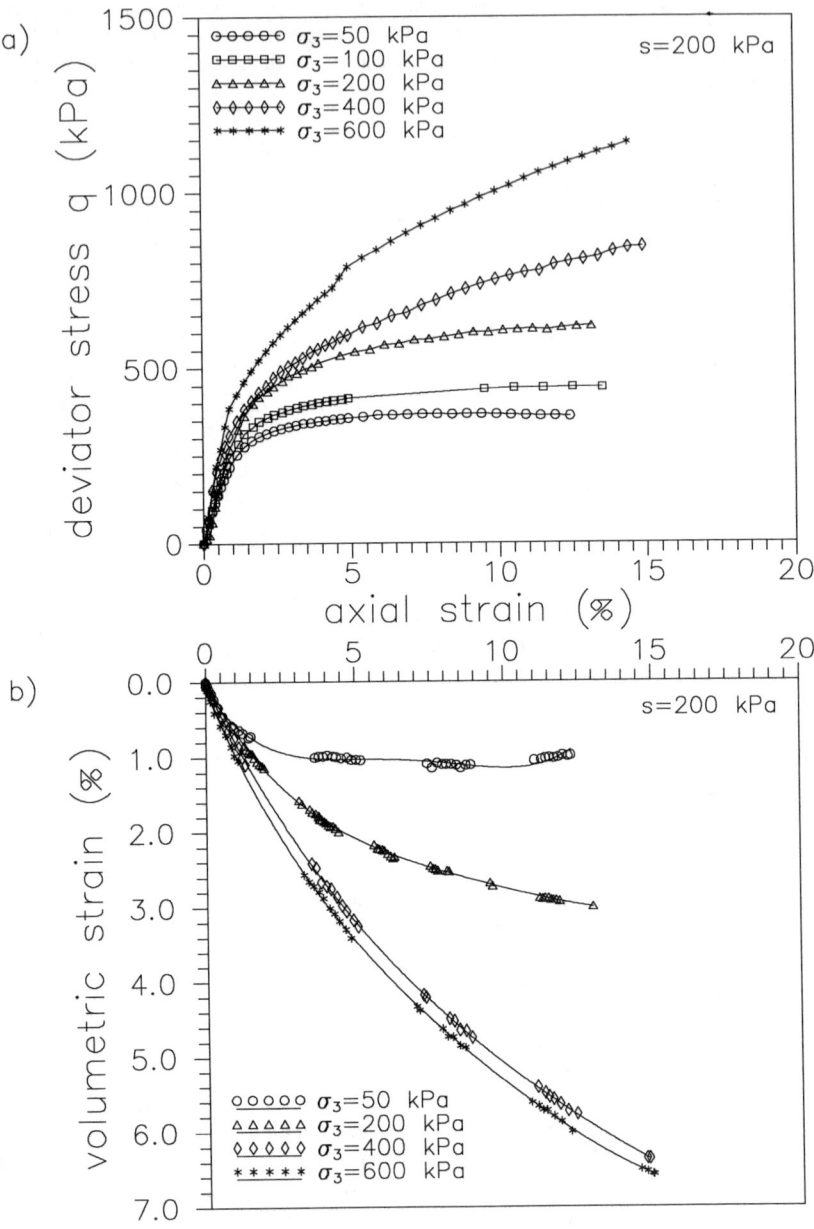

Figure 2 : Shear tests under a 200 kPa suction

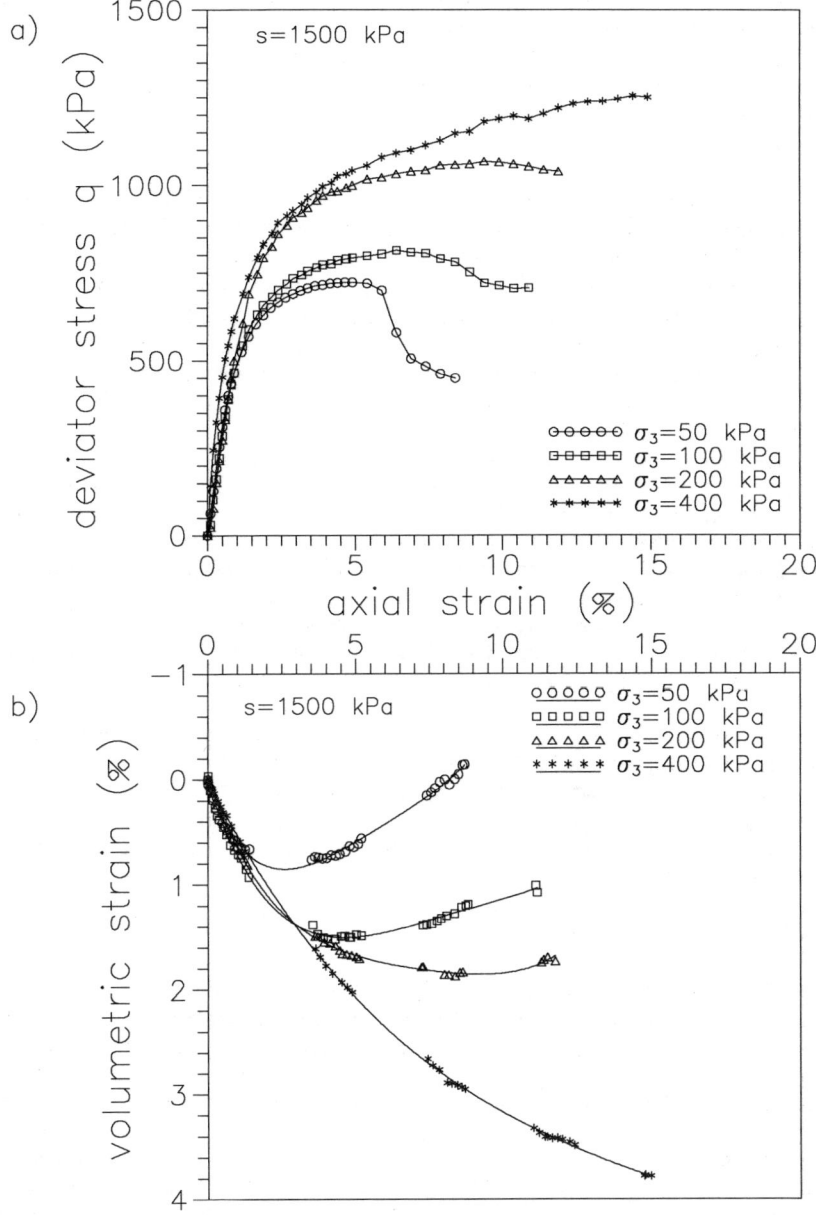

Figure 3 : Shear tests under a 1500 kPa suction

and of some softening at low confining stress ($\sigma_3 \leq 100$ kPa). Also the volume changes are quite different as compared to lower suction tests, with compressive-dilatant behavior observed for $\sigma_3 \leq 200$ kPa. If one compares these observations with some features of saturated soils behavior , one could conclude that suction has the same effect as density, since a high suction sample behaves like a high density soil, and vice-versa.

Figure 4 shows, for the same value of σ_3 (50 kPa), the effect of the suction on both shear stress and volume changes. There is a clear progressive increase of strength and stiffness with increasing suction ; post-peak softening behavior is observed at higher suctions (s $\geq$ 400 kPa). In these cases, localisation of the deformations and a well defined shear plane were observed. Under this low value of σ_3, compressive-dilatant behavior was observed at all values of suction, and the maximum observed compression decreases with increasing suction; also the maximum dilation is exhibited by the higher suction sample (s = 1500 kPa). The 0 kPa suction sample behaves somewhat differently than the others, with a much lower residual strength, and a change from compressive to dilatant behavior occurs at a smaller axial strain. These differences may be due to a hysteresis phenomenon. Since the initial suction of the compacted sample was estimated at 200 kPa by the filter paper method, only the 0 kPa suction sample has been wetted, whereas all the others have been dried. The hysteresis commonly observed in water retention curves is probably caused by changes in microstructure and could be responsible for changes in mechanical properties as well.. Such an hysteresis effect is well known on a hydraulic point of view when considering the water retention curves, and it should also be important on a mechanical point of view, since it probably involves some microstructure changes.

On figure 4, it is important to note that the effect of suction on volume changes is the opposite of the effect of stress, since the higher the suction, the higher the dilation. This is another limitation to the concept of effective stress, in triaxial conditions. At a higher confining stress (σ_3 = 400 kPa), there is no more evidence of dilating behavior, which shows that the effect of stress is greater than the effect of suction.

Interpretation of the volumetric behavior

The plotting of the volume changes during shear as a function of the logarithm of the mean stress p (p = ($\sigma_1 + \sigma_2 + \sigma_3$)/3) gives (figure 5) an explanation of the effect of suction on volume change behavior. It may be seen on the figure that the curves are similar to classical oedometer or isotropic compression curves, with two approximately linear parts, defining an "over consolidated" portion and a "normally consolidated" one, which should correspond to elastic and plastic behavior respectively. Whereas the initial slope of the curves appears more or less independent of suction, the "normally consolidated" portion has a slope which is suction dependent: the higher the suction, the smaller the compressibility; the apparent preconsolidation pressure also increases with suction; an exception is observed for the 400 kPa suction

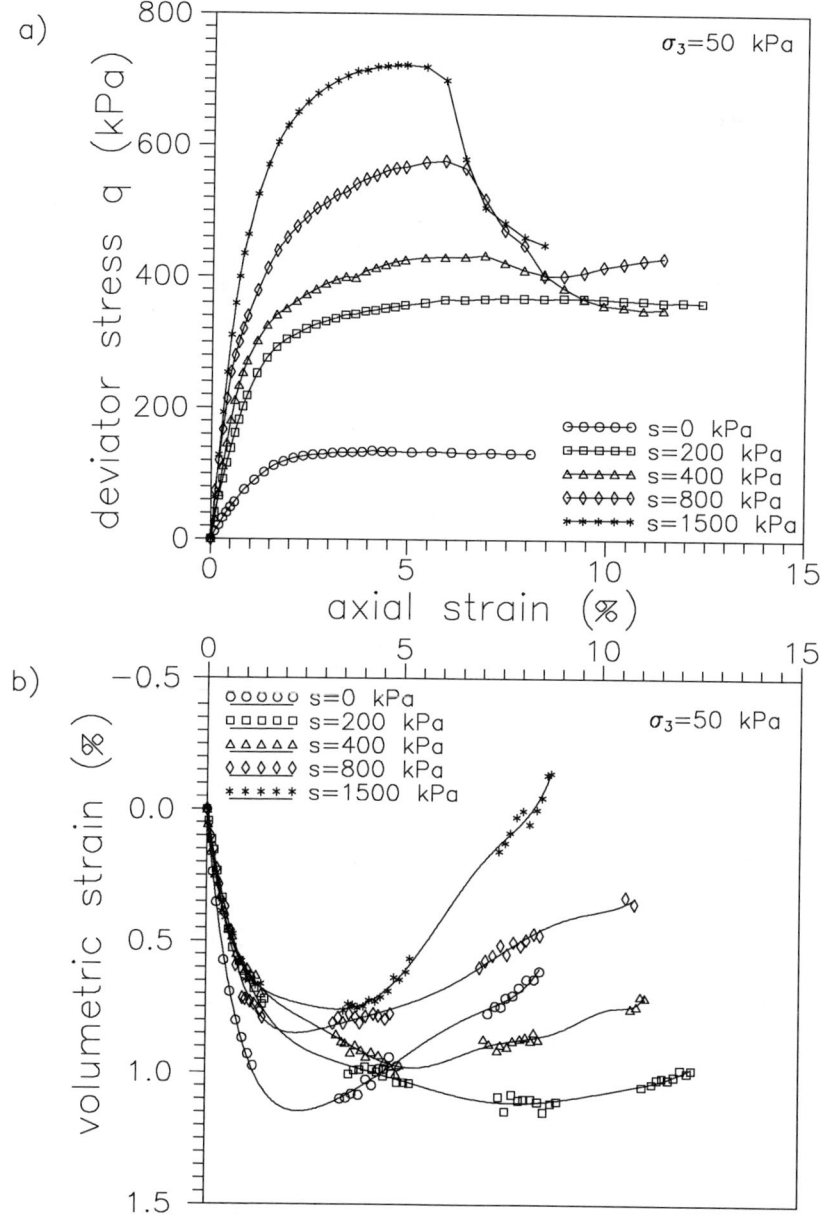

Figure 4 : Various constant suction tests made under a 50 kPa confining stress

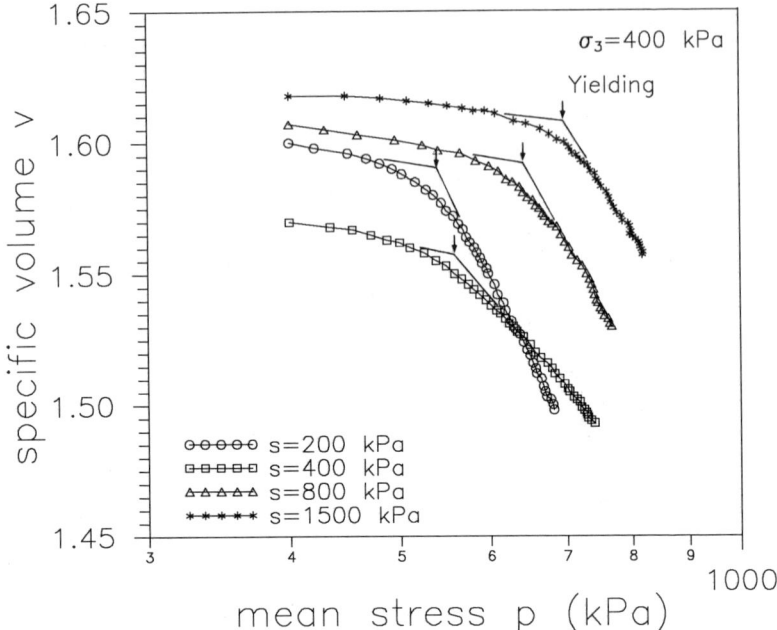

Figure 5 : Specific volume vs mean stress curves for various suctions,
under a 400 kPa confining stress

curve, probably because of a higher initial density, illustrated by the smaller initial volume. The same features were observed at other values of σ_3.

The curves of figure 5 are typical of the compressibility properties of unsaturated soils as a function of suction, as described in Alonso et al.(1987, 1990).

If one now reconsiders the volume change curves of figures 4 , one must remember that each of them corresponds to a different compressibility, which is a function of the applied suction; during shearing at a constant s_3 , the different volumetric compressibilities are mobilised by the increase of the mean stress, and their effect is apparently greater than the distortion volume changes generally accounted for when interpreting volume change behavior during shear. Here also, the analogy between suction and saturated soil density is valid, since suction, like density, reduces the compressibility of the sample; in other words, the various curves of figures 4 could have been obtained on a saturated soil, with tests made on soil samples prepared at various densities.

The deviatoric stresses corresponding to the apparent preconsolidation pressures of figure 5 have been reported on the stress-strain diagrams of figure 6, and it may be seen that, in each case, yielding appears at the same value of the axial strain, equal to approximately 2%.; below this value, the stress-strain curves are reasonably linear, with a steep slope; above 2%, either perfect plasticity or strain-hardening occurs, showing good correspondence between volumetric and shear behavior.

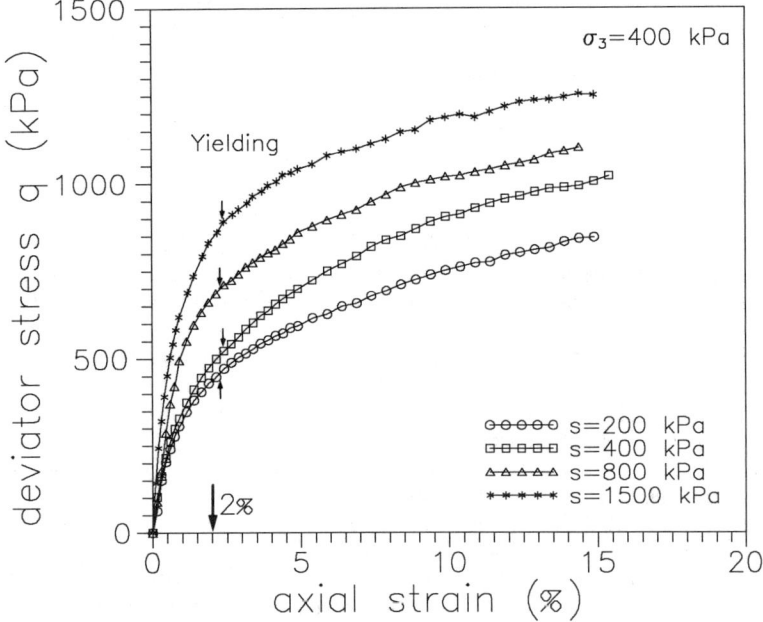

Figure 6 : Yielding points, according to the volumetric plastic criterion

CONCLUSION

In this series of osmotic suction-controlled triaxial compression tests, both stiffness and maximum shear strength were found to increase with increasing suction. Stress levels affect the compressive-dilatant behavior of the soil in the same manner as saturated soils. But suction affects the volume change of this unsaturated soil in the same way as density affects the behavior of saturated soils. Increasing suction has a significant influence on the compressibility of the soil, increasing the apparent preconsolidation pressure, or yield point, separating elastic from perfectly plastic or hardening behavior.

REFERENCES

1. ALONSO E.E., GENS A. & JOSA A. 1990. A constitutive model for partially saturated soils. Géotechnique 40(3), pp. 405-430.
2. ALONSO E.E., GENS A. & HIGHT D.W. 1987. Special problems soils. General report. Comptes Rendus de la 9e Conférence Européenne de Mécanique des Sols et des Travaux de Fondations, 3, pp. 1087-1146, Dublin.

3. BISHOP A.W. & DONALD I.B. 1961. The experimental study of partly saturated soil in the triaxial apparatus. Proceedings of the 5th International Conference on Soil Mechanics and Foundation Engineering, 1, pp. 13-21, Paris.

4. CUI Y.J. 1993. Etude du comportement d'un limon non-saturé, et de sa modélisation dans un cadre élasto-plastique. Thèse de l'Ecole Nationale des Ponts et Chaussées, Paris.

4. DELAGE P. & SURAJ DE SILVA G.P.R. 1992. Negative pore pressure and compacted soils. Mexican Society of Soil Mechanics. Raul Marsal Volume, pp. 219-232, Mexico.

5. DELAGE P., SURAJ DE SILVA G.P.R. & VICOL T. 1992. Suction controlled testing of non saturated soils with an osmotic consolidometer. 7th International Conference on Expansive Soils, Dallas, Texas.

6. DELAGE P., SURAJ DE SILVA G.P.R. & DE LAURE E. 1987. Un nouvel appareil triaxial pour les sols non saturés. Comptes Rendus de la 9ème Conférence Européenne de Mécanique des Sols et des Travaux de Fondations, 1, pp. 26-28, Dublin.

7. ESCARIO V. & SAEZ J. 1986. The shear strength of partly saturated soils. Géotechnique 36(3), pp. 453-456.

8. FREDLUND D.G. 1979 Appropriate concepts and technology for unsaturated soils. Canadian Geotechnical Journal 16, pp 121-136.

9. FREDLUND D.G., MORGENSTERN N.R. & WIDGER A. 1978. Shear strength of unsaturated soils. Canadian Geotechnical Journal 15, pp. 313-321.

10. GULHATI S.K. & SATIJA B.S. 1981. Shear strength of partially saturated soils. Proc. 10th. ICSMFE 1, pp. 609-612, Stockholm.

11. HO D.Y. & FREDLUND D.G. 1982. Strain rates for unsaturated soil shear strength testing. Proceedings of the 7th South-east Asian Geotechnical Conference, pp. 787-803, Hong-Kong.

12. KASSIF G. & BEN SHALOM A. 1971. Experimental relationship between swell pressure and suction. Géotechnique 21(3), pp. 245-255.

13. KOMORNIK A., LIVNEH M. & SMUCHA S. 1980. Shear strength and swelling of clays under suction. 4th International Conference on Engineering Science, 1, pp. 206-226, Denver.

15. SURAJ DE SILVA G.P.R. 1987. Etude expérimentale du comportement d'un limon non saturé sous succion contrôlée. Thèse de l'Ecole Nationale des Ponts et Chaussées, Paris, 222 p.

16. TOLL D.G. 1990. A framework for unsaturated soil behavior. Géotechnique 40, No.1, pp. 31-44.

17. WHEELER S. J. 1991. Technical note: An alternative framework for unsaturated soil behavior. Géotechnique 41(2), pp. 257-261.

18. WHEELER S.J. & SIVAKUMAR V. 1992. Critical state concepts for unsaturated soil. Proceedings of the 7th International Conference on Expansive Soils. Dallas, Texas 1, pp. 167-172.

19. WILLIAMS J. & SHAYKEWICH C.F. 1969. An evaluation of polyethylene glycol (PEG) 6000 and PEG 20000 in the osmotic control of soil water matric potential. Canadian Journal of Soil Science 102(6), pp. 394-398.

Elasto-Plastic Volume Change of Unsaturated Compacted Clay

V. Sivakumar[1] and S.J. Wheeler[1]

Abstract

Isotropic compression tests were performed on samples of unsaturated compacted kaolin in a modified triaxial cell that allowed separate control of pore water pressure and pore air pressure. Each test began with an equalization stage, during which the suction was reduced to a pre-selected value while maintaining the mean net stress constant. This was followed by a ramped consolidation stage, in which the mean net stress was increased at constant suction. The variation of specific volume during ramped consolidation showed clear evidence of a yield point and an elasto-plastic form of behaviour. The complex pattern of swelling and collapse observed during equalization was also explained by elasto-plastic behaviour and the existence of a yield curve in suction versus mean net stress space. When loaded to virgin conditions the soil states were represented by an isotropic normal compression hyper-line, defined by an expression relating specific volume to the mean net stress and suction.

Introduction

It is now generally accepted that the mechanical behaviour of unsaturated soil cannot be related to a single "effective stress" and that total stress σ, pore air pressure u_a and pore water pressure u_w must be combined in two independent stress parameters (Bishop and Blight, 1963; Fredlund and Morgenstern, 1977). The two stress parameters normally selected are the "net stress" $\sigma - u_a$ and the matrix suction $u_a - u_w$ (referred to hereafter simply as "suction").

This paper is concerned with the volume change behaviour of compacted kaolin caused by changes of net stress and suction under isotropic loading conditions. This aspect of unsaturated soil behaviour is set within a generalised

[1]Research Assistant and University Lecturer respectively, Department of Engineering Science, University of Oxford, Parks Road, Oxford OX1 3PJ, UK.

elasto-plastic stress-strain framework based on critical state concepts.

Elasto-Plastic Framework

Alonso, Gens and Josa (1990) proposed an elasto-plastic critical state framework for unsaturated soil involving four state variables : mean net stress p, deviator stress q, suction s and specific volume v, defined as follows (for the simplified axisymmetric conditions of the triaxial test, $\sigma_2 = \sigma_3$)

$$p \quad = \quad (\sigma_1 + 2\sigma_3)/3 - u_a \tag{1}$$

$$q \quad = \quad \sigma_1 - \sigma_3 \tag{2}$$

$$s \quad = \quad u_a - u_w \tag{3}$$

$$v \quad = \quad 1 + e \tag{4}$$

The unsaturated critical state framework therefore involves one more stress state variable than saturated critical state models (because of the need to include net stresses and suction instead of just effective stresses). In two earlier papers (Wheeler and Sivakumar, 1992, 1993) the authors employed water content w as a fifth state variable, but more recent evidence (Wheeler and Sivakumar, in press) suggests that perhaps water content should not be treated as a state variable.

This paper is restricted to isotropic loading conditions, when the deviator stress q is zero. Included within the proposed critical state framework is an isotropic normal compression "hyper-line", representing soil states when isotropically loaded to virgin conditions. The term "hyper-line" was coined to describe a locus of states defined in four-dimensional mathematical space by two independent equations. The two equations defining the isotropic normal compression hyper-line take the general form

$$q \quad = \quad 0 \tag{5}$$

$$v \quad = \quad f(p, s) \tag{6}$$

Possible forms for an expression relating specific volume v to mean net stress p and suction s, as shown in Equation 6, have been proposed by several previous authors, including Fredlund and Morgenstern (1976) and Lloret and Alonso (1985). A fundamental feature of the proposed critical state framework is however that movement along the isotropic normal compression hyper-line corresponds to plastic (irrecoverable) compression and expansion of a yield curve plotted in s versus p space (part of a yield surface in p, q, s space). Soil states will, therefore, lie on the isotropic normal compression hyper-line only if the soil is loaded to a virgin condition. Unloading, due to a reduction of mean net stress p or an appropriate change of suction s, will result in elastic volume changes as the stress point moves back inside the yield curve, and the soil state will no longer lie on the isotropic normal compression hyper-line. Re-loading will occur elastically until the yield curve is reached once more and the soil state returns to the isotropic normal compression hyper-line.

Experimental Procedure

A series of 29 controlled suction isotropic compression tests was performed on samples of unsaturated compacted kaolin, in order to examine further the volume change behaviour of unsaturated soil within the context of the proposed elasto-plastic framework. The test programme is described in detail by Sivakumar (1993).

Triaxial samples of unsaturated compacted speswhite kaolin, 50 mm in diameter and 100 mm high, were prepared by compaction in a mould at a water content of 25% (4% less than optimum). All samples were compacted in 9 layers, with each layer "statically" compacted in a compression frame at a fixed displacement rate of 1.5 mm/min to a vertical total stress of 400 kPa. This procedure corresponded to a compactive effort considerably less than the various standard dynamic compaction tests and resulted in a dry density of 1.20 Mg/m^3, a specific volume of 2.21 and a degree of saturation of about 54%. The intention was to produce samples with a low value of yield stress, so that it was relatively easy subsequently to consolidate them to a virgin state. All 29 samples were compacted in an identical fashion, in order to produce the same initial soil fabric in every test.

The tests were conducted in two double-walled triaxial cells designed for testing unsaturated soil samples (Wheeler, 1988). Pore water pressure u_w was applied at the base of the sample via a porous filter with an air entry value of 500 kPa. Values of pore water pressure were maintained above atmospheric, using the "axis translation" principle first proposed by Hilf (1956). Pore air pressure u_a was applied at the top of the sample via a filter with a low air entry value. The cell pressure σ_3, pore water pressure u_w and pore air pressure u_a were each controlled independently by stepper motors operating regulators on a compressed air supply. The stepper motors were operated by a computerised control and logging system which enabled any required stress path to be followed.

Changes of sample volume were measured by monitoring the flow of water into the inner chamber of the double-walled cell. A rolling diaphragm seal on the loading ram prevented leakage of water from the cell, and the system was calibrated for apparent volume changes caused by compression of the water within the cell, absorption of water by the acrylic cell wall and expansion of the cell with pressure (this last effect was minimised by the double-walled construction). A second volume change unit on the drainage line from the base of the sample was used to measure the volume of water flowing in or out of the sample.

The first phase of each test, after setting up the sample in the triaxial cell, consisted of an equalization stage, during which the sample was brought to a mean net stress p of 50 kPa and a suction of 100, 200 or 300 kPa. This was achieved by applying a water back pressure of 50 kPa to the base of the sample, an air back pressure of 150, 250 or 350 kPa to the top of the sample and a cell pressure of 200, 300 or 400 kPa respectively. Five samples were prepared at a suction of 100 kPa, eleven at a suction of 200 kPa and four at a suction of 300 kPa.

Eight further compacted samples were brought to a saturated state at zero suction. In order to achieve full saturation of these samples during the equalization stage, the air back pressure at the top of the sample was temporarily reduced to 35 kPa (15 kPa less than the water back pressure applied at the base of the sample), resulting in a steady upward flow of water through the sample by the end of the stage. An additional slight difference in procedure for the samples prepared to zero suction was that the value of p during equalization was 40 kPa, rather than the standard value of 50 kPa used in all other tests. The reason for choosing this lower value of p was to try to ensure (unsuccessfully as it transpired) that the yield stress induced by the initial compaction process could be observed in the subsequent consolidation stage, bearing in mind that the yield stress would be lowest at zero suction (see below).

The equalization stage of each test was terminated when the rate of change of water content fell to less than 0.04% per day (typically after 6 to 10 days).

Following equalization, each sample was isotropically consolidated to a pre-selected value of p (typically 100, 150, 200, 250 or 300 kPa) by ramping the cell pressure at a rate of 0.6 kPa/hour while holding constant both water back pressure and air back pressure (thus maintaining constant suction). The rate of increase of cell pressure was selected to give acceptable dissipation of excess pore water pressure throughout the sample during ramped consolidation (see Sivakumar, 1993). On reaching the target value of p, a rest period of 24 hours was allowed, to ensure complete dissipation of any excess pore water pressure. The entire ramped consolidation stage took 5 to 18 days, depending upon the target value of p.

Fig. 1 shows the stress paths, plotted as suction s versus mean net stress p, for the equalization and ramped consolidation stages. On initial setting up in the triaxial cell each sample was at point A, with the cell pressure and pore air pressure both zero (resulting in $p = 0$) and a large negative value of pore water pressure (a high value of suction) produced by the compaction process. On the

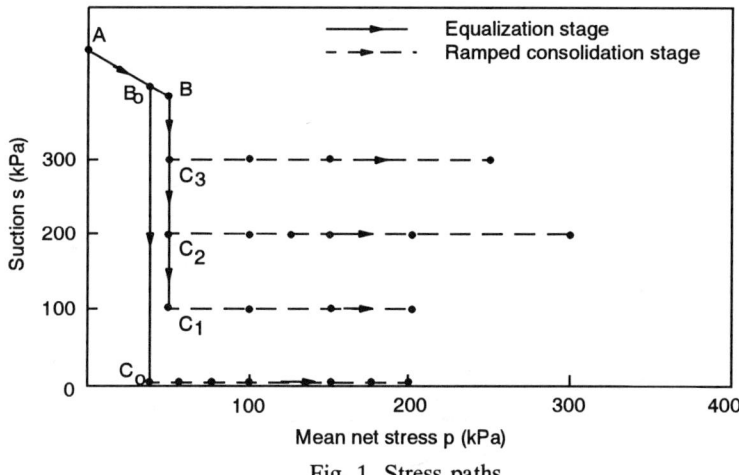

Fig. 1 Stress paths

application of the increments of cell pressure, air back pressure and water back pressure at the start of the equalization stage the soil state moved to point B (or B_0 for the samples to be prepared to zero suction). At B or B_0 the pore air pressure u_a throughout the sample had equalized at the applied air back pressure value, because of the high value of air permeability, but the pore water pressure u_w had not yet risen to the applied water back pressure value, because of the relatively low value of water permeability.   During the subsequent equalization stage the soil state moved to point C_0, C_1, C_2 or C_3, with suction s falling (wetting) while mean net stress p remained constant.   Finally, in the ramped consolidation stage, p was increased to a value of 100 to 300 kPa while s remained constant.

Behaviour during Wetting

Fig. 2 shows the change of specific volume that occurred during the equalization stage plotted against time.   Each test started with an "immediate" reduction in specific volume of about 0.01, corresponding to the rapid increment of mean net stress p from zero to 50 kPa (or 40 kPa for the samples to be brought to zero suction).   This initial compression was followed in all tests by an increase in specific volume (of 0.015 to 0.03), as swelling took place due to the reduction in suction during equalization.   In tests with an applied suction of zero or 100 kPa the swelling phase was followed by a reduction in volume (collapse) during the later part of equalization.   This collapse phase was a far more significant in the tests at zero suction than those at 100 kPa (the reductions of specific volume were about 0.08 and 0.01 respectively), and there was no evidence of collapse in the tests at suctions of 200 and 300 kPa.

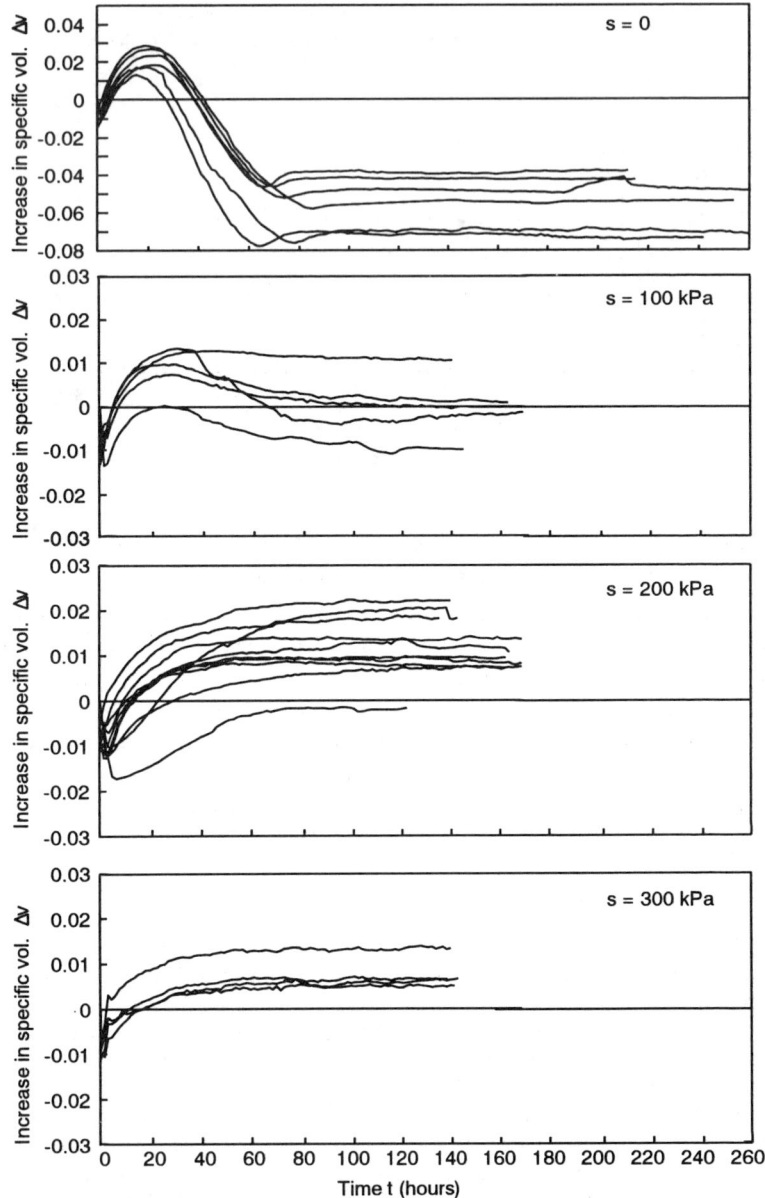

Fig. 2 Change of specific volume during equalization stage

The behaviour observed during the equalization stage can be explained by the existence of a yield curve in s, p space, of the shape proposed by Alonso, Gens, and Josa (1990). Fig. 3 shows the stress paths adopted during the equalization stage and the approximate position of the yield curve produced by the initial compaction process. Inside the yield curve the soil behaviour would be elastic, with an increase in p causing elastic compression and a decrease in s causing elastic swelling. If the yield curve were reached, a further increase in p or reduction in s would lead to expansion of the yield curve and a large component of plastic compression (collapse). Inspection of Fig. 2 suggests that the yield curve in Fig. 3 passed outside points C_2 and C_3 (because no collapse occurred in the tests at s = 200 kPa or s = 300 kPa), just inside point C_1 (because a small amount of collapse was observed in tests with s = 100 kPa) and well inside point C_0 (because there was a large amount of collapse in tests with s = 0).

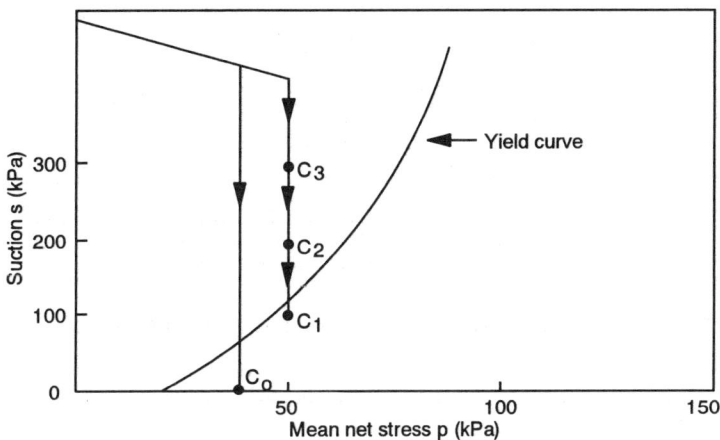

Fig. 3   Yield curve in s versus p space

Behaviour during Isotropic Loading

The variation of specific volume v with mean net stress p during the ramped consolidation stage (with p on a logarithmic scale) is shown in Fig. 4. The curves presented in Fig. 4 are not quite perfect constant suction compression plots, because of slight non-equalization of pore water pressure u_w throughout the samples during ramped consolidation. This point is demonstrated by the small vertical step at the end of each curve, corresponding to complete equalization of pore water pressure to the water back pressure value during the 24 hour rest period at the end of ramped consolidation.

For each test shown in Fig. 4 there was a clear yield point, identified by a
marked change in slope of the continuous plot of specific volume versus logarithm
of mean net stress, and a yield value of p was calculated by using Casagrande's
graphical construction. These values of yield stress, which showed very little
variation (about ± 5 kPa) at a given value of suction, are plotted in Fig. 5. The

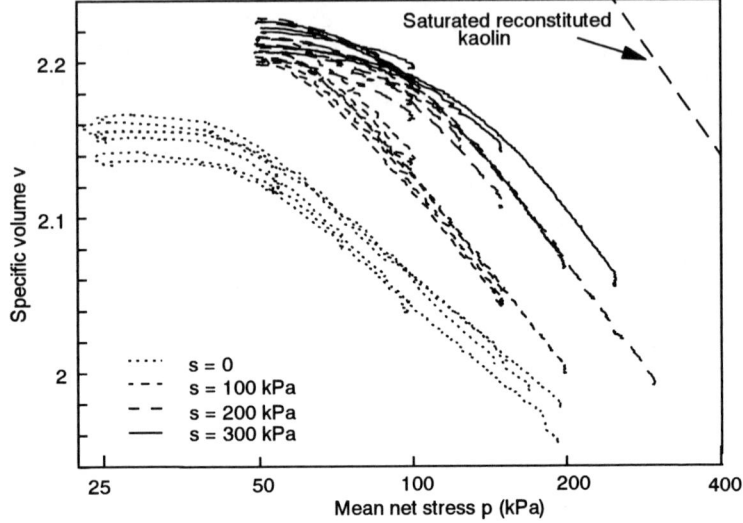

Fig. 4 Variation of specific volume during ramped consolidation

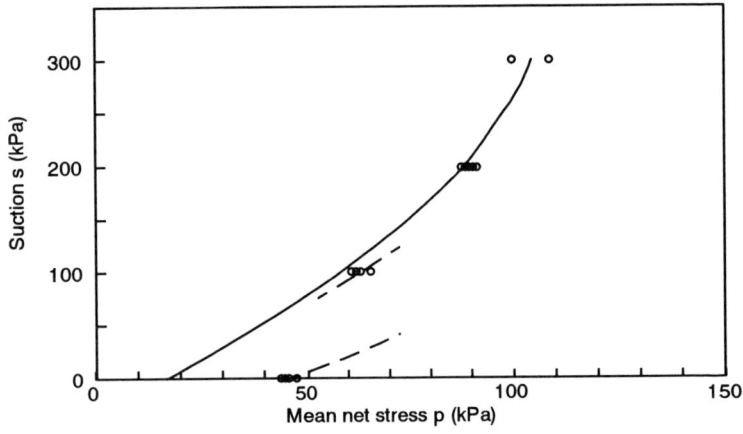

Fig. 5 Yield points observed during ramped consolidation

approximate position of the yield curve produced by the initial compaction process has been sketched as a solid line in Fig. 5. This initial yield curve passes though the yield points at suctions of 200 kPa and 300 kPa, because there was no evidence of collapse during the equalization stage of these tests (suggesting that there was no expansion of the yield curve during equalization). The initial yield curve was however drawn inside the yield points identified during ramped consolidation at suctions of zero and 100 kPa, because the collapse that occurred during equalization in these tests suggested corresponding expansions of the curve (represented by the dashed lines in Fig. 5). As a consequence the values of yield stress that would have been expected in the tests at $s = 0$ and $s = 100$ kPa were 40 kPa and 50 kPa respectively (the values of p applied during the equalization stage). Inspection of Fig. 5 suggests therefore that the use of Casagrande's graphical procedure may have led to slight over-estimation of the yield stress or that the apparent yield point was affected by the change of stress path direction.

Further inspection of the ramped consolidation plots shown in Fig. 4 indicates that when the yield stress at a particular value of suction was exceeded, the soil state fell on an isotropic normal compression hyper-line defined by a linear relationship

$$v = N(s) - \lambda(s)ln\left(\frac{p}{p_{at}}\right) \tag{7}$$

To make the expression dimensionally consistent, and also to minimise any error in the evaluation of $N(s)$, atmospheric pressure p_{at} (taken as 100 kPa) has been introduced in Equation 7.

Both the slope $\lambda(s)$ and intercept $N(s)$ of the isotropic normal compression hyper-line, taken from Fig. 4, were found to be functions of suction, as shown in Fig. 6. The intercept $N(s)$ increased with increasing suction, consistent with expected behaviour. The slope $\lambda(s)$ showed relatively little variation with s for suctions between 100 and 300 kPa but a significant drop when s was reduced to zero. This variation of $\lambda(s)$ was not consistent with the proposals of Alonso, Gens and Josa (1990) who proposed a monotonic decrease of $\lambda(s)$ with increasing suction, so that the normal compression lines for different values of suction diverged with increasing p. The behaviour illustrated in Fig. 4 is however more consistent with a modified model (Josa, Balmaceda, Gens and Alonso, 1992), which predicted that normal compression lines for different values of suction would start to converge once a critical value of p (corresponding to the maximum collapse on wetting) was exceeded.

The normal compression line for zero suction (saturated conditions) fell considerably below the corresponding normal compression line for saturated reconstituted kaolin samples (prepared from a slurry) and the slope was also

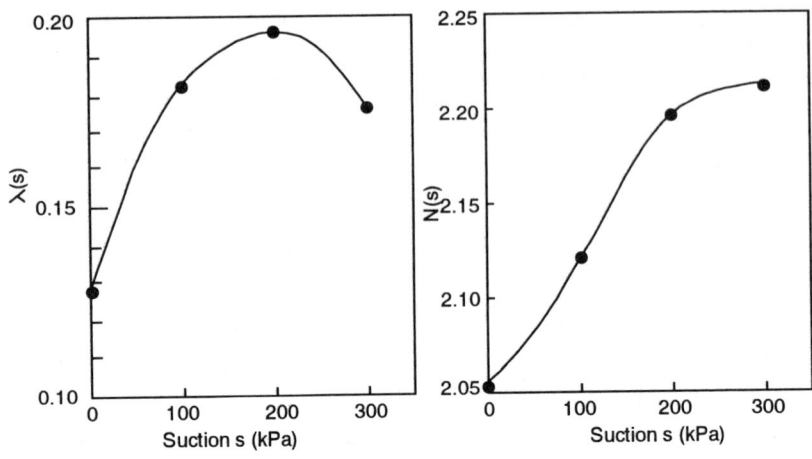

Fig. 6   Variation of $\lambda(s)$ and $N(s)$ with suction

considerably less (see Fig. 4). This difference between saturated compacted kaolin
and saturated reconstituted kaolin indicated the significance of the soil fabric
produced by the compaction process.

Discussion and Conclusions

The variation of specific volume v caused by increasing mean net stress p
during the ramped consolidation stages clearly showed the occurrence of yielding
and an elasto-plastic form of behaviour (see Fig. 4). The complex pattern of
swelling and collapse caused by reducing suction s during the preceding
equalization stages (see Fig. 2) was also explained by yielding and elasto-plastic
behaviour.   When loaded to virgin states, by monotonically increasing p or
decreasing s, the soil state reached a unique normal compression hyper-line,
defined by Equation 7. On unloading (by decreasing p or increasing s) it is likely
that the soil response would be essentially elastic, with changes in v related to
changes in p and s by two elastic swelling indices κ and κ_s (see Alonso, Gens and
Josa, 1990).

The shape of the yield curve in s versus p space has significant practical
implications for the selection of a suitable experimental procedure for compression
testing of unsaturated samples.   Fig. 7 shows the stress path for a point in the
interior of a soil sample subjected to a step increment of total stress followed by
a pore pressure dissipation phase.   As the increment of total stress is applied,
excess pore air pressure and excess pore water pressure will be generated within
the sample. Any excess pore air pressure dissipates very quickly, whereas excess

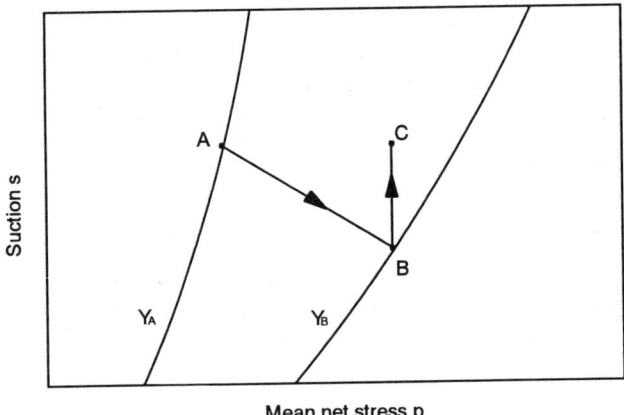

Mean net stress p

Fig. 7 Stress path for step-loading consolidation test

pore water pressure takes much longer to dissipate. As a consequence, the stress state moves initially from point A to point B (at the final value of p but a reduced value of s) and the stress state then slowly moves from point B to the final point C as the dissipation of excess pore water pressure leads to a corresponding increase of suction. The result of this stress path is that the yield curve expands from an initial position Y_A to a position Y_B during stress path AB, but the yield curve then remains at position Y_B during stress path BC (which corresponds to elastic unloading inside the yield curve). The soil sample is therefore not in a virgin state at the end of the consolidation phase, and greater plastic compression will have occurred than would have been the case on the desired stress path direct from A to C. This explains why Barden, Madedor and Sides (1969) found that the compression observed during step-loading oedometer tests on unsaturated samples increased with increasing stress increment ratio. A better way of conducting consolidation tests on unsaturated samples would therefore be to ramp the total stress at a rate that was sufficiently slow to ensure that the excess pore water pressure remained acceptably small. This was the procedure employed in the test programme described in this paper.

The position and shape of the normal compression hyper-line, defined by Equation 7, could be dependent upon the initial soil fabric produced by the compaction process. This point is highlighted by the difference between saturated compacted kaolin ($s=0$) and saturated reconstituted kaolin (see Fig. 4). All samples tested in the experimental programme reported here were compacted in an identical fashion and were therefore expected to have the same initial fabric. Samples produced by compaction at a different water content or with a different compactive effort would have a different fabric and therefore the soil parameters $N(s)$ and $\lambda(s)$ would not necessarily be expected to take the same values or show the same variation with suction. It might therefore be necessary to treat samples produced by different compaction procedures as different soils.

Acknowledgements

The experimental research described in the paper was performed by the authors in the Department of Civil and Structural Engineering at the University of Sheffield. Financial support was provided by the UK Science and Engineering Research Council.

References

Alonso, E.E., Gens, A. and Josa, A. (1990). A constitutive model for partially saturated soils. Géotechnique, 40(3), 405-430.

Barden, L., Madedor, A.O. and Sides, G.R. (1969). Volume change characteristics of unsaturated clay. J. of the Soil Mechanics and Foundations Division, Proc. ASCE, 95 (SM1), 35-51.

Bishop, A.W. and Blight, G.E. (1963). Some aspects of effective stress in saturated and partly saturated soils. Géotechnique, 13(3), 177-197.

Fredlund, D.G. and Morgenstern, N.R. (1976). Constitutive relations for volume change in unsaturated soils. Canadian Geotechnical J., 13, 261-276.

Fredlund, D.G. and Morgenstern, N.R. (1977). Stress state variables for unsaturated soils. Proc. ASCE, 103 (GT5), 447-466.

Hilf, J.W. (1956). An investigation of pore water pressure in compacted cohesive soils. US Bureau of Reclamation, Tech. Memo. 654, Denver.

Josa, A., Balmaceda, A., Gens, A. and Alonso, E.E. (1992). An elasto-plastic model for partially saturated soils exhibiting a maximum of collapse. Proc. 3rd Int. Conf. on Computational Plasticity, Barcelona, 1, 815-826.

Lloret, A. and Alonso, E.E. (1985). State surfaces for partially saturated soils. Proc. 11th ICSMFE, San Francisco, 2, 557-562.

Sivakumar, V. (1993). A critical state framework for unsaturated soil. Phd thesis, University of Sheffield.

Wheeler, S.J. (1988). The undrained shear strength of soils containing large gas bubbles. Géotechnique, 38 (3), 399-413.

Wheeler, S.J. and Sivakumar, V. (1992). Critical state concepts for unsaturated soil. Proc. 7th Int. Conf. on Expansive Soils, Dallas, 1, 167-172.

Wheeler, S.J. and Sivakumar, V. (1993). Development and application of a critical state model for unsaturated soil. Proc. Wroth Memorial Symposium on Predictive Soil Mechanics, Oxford.

Wheeler, S.J. and Sivakumar, V. (in press). A critical state framework for unsaturated soil. Submitted to Géotechnique.

One and Three Dimensional, Three Phase Deformation in Soil

Thomas V. Edgar, M. ASCE[1]

Abstract

An initially saturated soil exhibits two distinct deformation patterns as it drys. First, it deforms vertically until the soil begins to drain. Secondly, it deforms laterally as well as vertically as the pores drain. This paper presents a simple equilibrium thermodynamic analysis for the appropriate stress state variables and the associated strains. These stress state variables are based on macroscopic stresses which are easily determined. This analysis is extended to develop a constitutive relation which provides a continuous transition from saturated to unsaturated conditions.

Introduction

Terzaghi established the effective stress concept for saturated soils in 1925 (Terzaghi, 1925) and its use is one of the foundations of the field of soil mechanics. The concept states that soils behave in response to changes in a stress state variable, the effective stress, defined as the difference between the total stress σ applied to a soil and the pore pressure (usually the water pressure, u_w but possibly the air pressure, u_a) acting inside the soil. Deformation, soil strength and bearing capacity are all related to the effective stress acting on the soil.

The concept is ill defined when the soil is partially saturated, that is, when both water and air exist jointly in the pore space. Each fluid has a different pressure, and the effects of these pressures on soil behavior are altered by the fabric of the soil.

Several techniques currently exist for dealing with this problem (Fredlund and Morgenstern, 1977). The most common method uses Bishop's χ (chi) parameter

---

[1]Associate Professor, Department of Civil and Architectural Engineering, University of Wyoming, Laramie, WY   82071

(Bishop, 1959), a function of the degree of saturation of the soil. An effective stress is then defined by

$$\overline{\sigma} = (\sigma - u_a) - \chi(u_a - u_w) \ . \tag{1}$$

While this equation is fairly easy to use, it has two significant limitations. First, χ is not a unique variable. It is a function of the degree of saturation, the past stress history and the deformation. Secondly, and more fundamentally incorrect, it uses a material property to define a state of stress. Accepted practice in mechanics separates the state of stress (the summation of forces) from the material properties (the constitutive properties).

Fredlund and Morgenstern (1977) have defined three significant stress state variables (only two of which are independent) based on a force balance acting on a soil having solid, water and air phases. The stress state variables are:

$$(\sigma - u_a), \ (\sigma - u_w) \ \text{and} \ (u_a - u_w) \ . \tag{2}$$

These variables are then used to determine the constitutive relationships acting between the state of stress and the deformation or strength.

This paper develops an alternative development of these stress state variables, presents a new constitutive model for one-dimensional saturated-nonsaturated consolidation and indicates research being performed at the University of Wyoming to determine its appropriateness.

The Energy Equation

An initially saturated soil exhibits two distinct deformation patterns as it drys. As the water content decreases, the saturated soil deforms in the vertical direction. Increasing capillary stress causes the soil to dry at the surface and crack in the lateral directions. Cracking can occur while the soil clods are saturated or draining.

The open soil system shown in Fig. 1 has three phases, a solid, a liquid (water) and a gas (air). The control volume for this system maintains a constant volume of solids, while the void phases may pass through the system boundary in response to changes in fluid pressures as indicated by the differential volumes in the figure (Edgar, 1983). If the pressure changes are low, the densities of the phases can be assumed constant, even though the total volume may change as the water and air flow through the boundaries. This is similar to the soil element commonly used to describe consolidation.

The increase in internal energy in this system dE is caused by heat flowing into the system, dQ, or by work being applied to it, dW, such that (Haase, 1969)

$$dE = dQ + dW \quad . \tag{3}$$

If the system is adiabatic, then $dQ = 0$, and the energy change is caused solely by work being performed on it. One form of work is that due to the application of a total stress to the system. Stress can be in the form of normal stresses σ_{ii} applied perpendicular to or shear stresses $\sigma_{ij, i \neq j}$ applied parallel to the surface of the control volume. In this case, the work is

$$dW_{tot} = -V_o \sum_{i=1}^{3} \sum_{j=1}^{3} \sigma_{ij} d\varepsilon_{ij} \tag{4}$$

where V_o is the undeformed volume and σ_{ij} and ε_{ij} are the stress and strain tensors, respectively (Haase, 1969). For one dimensional loading in which no lateral deformations occur (i.e., K_o conditions), then

$$\varepsilon_{22} = \varepsilon_{33} = 0 \tag{5}$$

and

$$\sigma_{12} = \sigma_{13} = \sigma_{23} = 0. \tag{6}$$

This reduces the work equation to

$$dW_{tot} = -\sigma_{11} V_o d\varepsilon_{11}. \tag{7}$$

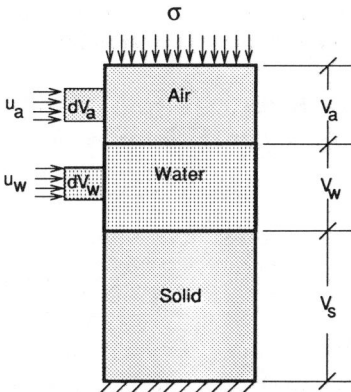

Figure 1 - Control Volume with Constant Volume of Solids
and Variable Volumes of Water and Air

Because

$$dV = V_o d\varepsilon_{11} \tag{8}$$

is the 1-D volumetric strain, then the work performed by this one dimensional system is

$$dW_{tot} = -\sigma \, dV \tag{9}$$

The sign of this equation indicates that positive compression (the soil mechanics convention) doing positive work results in a volume decrease.

The total work expression given in Eq. 9 has the same appearance as the work performed by a piston acting on a homogeneous fluid, the "$P \, dV$" work. This relationship is very different because of the limitations on the stress and strain imposed because the soil is able to resist shear stress.

Water and air can enter the system by being pushed through the boundary walls due to fluid pressure changes. This is work performed on a homogeneous fluid and is described by

$$dW_w = u_w \, dV_w \tag{10}$$

and

$$dW_a = u_a \, dV_a \tag{11}$$

respectively. In these cases, an increase in fluid pressure causes an increase in volume, therefore the sign for positive work performed on the system is positive.

In addition, the added masses (or volume) of water m_w and air m_a have their own associated internal energy (referred to as the sensible energy) equal to

$$dE_w = e_w dm_w = e_w \rho_w dV_w \tag{12}$$

and

$$dE_a = e_a dm_a = e_a \rho_a dV_a. \tag{13}$$

where ρ represents the mass density of the phase. All these terms may be combined to determine the total energy change in the soil,

$$dE = -\sigma\,dV + u_w dV_w + u_a dV_a + e_w\rho_w dV_w + e_a\rho_a dV_a.\qquad(14)$$

This describes the relationship between change in total energy of a soil and the application of total stress, water pressure and air pressure. Notice that the interfacial tension between the water and air is not included. That interaction occurs inside the element and is not considered individually as a system. The energy expended in deforming the interface between the water and air is automatically reflected in the different pressures of the two fluids. Conversely, interfacial tension would be required if the energy of each individual phase was being considered.

This equation can be modified by considering the relationship between the phase and total volumes. The total volume is equal to the sum of the phase volumes,

$$V = V_s + V_w + V_a.\qquad(15)$$

The deformation of the total volume is equal to the sum of the deformations of the phase volumes,

$$dV = dV_s + dV_w + dV_a,\qquad(16)$$

but the volume of solid in the element is constant. Hence the deformation occurs only in the void volumes,

$$dV = dV_v = dV_w + dV_a.\qquad(17)$$

These values in the three permutations may be substituted into the energy equation to give the stress state variables defined by Fredlund and Morgenstern. In addition, it also indicates the appropriate extensive variable associated with the stresses. For example, solving Eq. 17 for dV_w and substituting in Eq. 14 produces

$$dE = -(\sigma - u_w)dV_v + (u_a - u_w)dV_a\qquad(18)$$

where the sensible heats of the water and air have been neglected because they are small compared to the other terms. The reference volume has a constant volume of solid, so that if both sides are divided by V_s, then Eq. 18 may be written

$$\frac{dE}{V_s} = -(\sigma - u_w)de + (u_a - u_w)de_a\qquad(19)$$

where e is the void ratio (V_v/V_s) and e_a is the air void ratio (V_a/V_s). The first term (the traditionally defined effective stress multiplied by the change in void ratio) represents the area under a standard consolidation curve (Fig. 2a) whereas the second

term (the product of the capillary stress and the change in air void ratio) is related to
the moisture characteristic curve shown in Fig. 2b. This figure is shown in void ratio
terms and indicates by the slope of the dark line that the soil is deforming. Figure 2c
shows just the change in the air void ratio as a function of capillary stress. It also
shows the air void ratio is zero below the displacement pressure, p_d.

Solving Eq. 17 for dV and dV_a produces the following equivalent expressions

$$\frac{dE}{V_s} = -(\sigma - u_w)de_w + (\sigma - u_a)de_a \quad \text{and} \quad \frac{dE}{V_s} = -(\sigma - u_a)de - (u_a - u_w)de_w \; , \quad (20)$$

respectively.

An important aspect of Eq. 19 is that it indicates if the soil remains saturated
(such $de_a = 0$), the capillary stress does not influence the energy of the system. The
implications of this can be investigated in a standard pressure plate used to determine
the moisture characteristic curve for a soil. A saturated sample is placed on a high air
entry pressure plate and the air pressure is increased while the water pressure is
maintained at atmospheric pressure. Initially, the water drains from the soil while the
soil remains saturated. The changing air pressure (and not the capillary stress) causes
the volume to change. Physically, the surface tension of the air-water interface bridges
the void space between the particles while the soil is saturated as shown in Fig 3a.
Figure 3b shows the contractile skin being acted on by the air pressure. The air
pressure indirectly compresses the soil by forcing the interface to compress the particle
structure just as the total stress does by acting directly on the particles. Therefore, the
air pressure must be acting like a total stress on the sample, at least in the one
dimensional case considered here. Figure 3c indicates the vertical stress components
acting on the air-water interface.

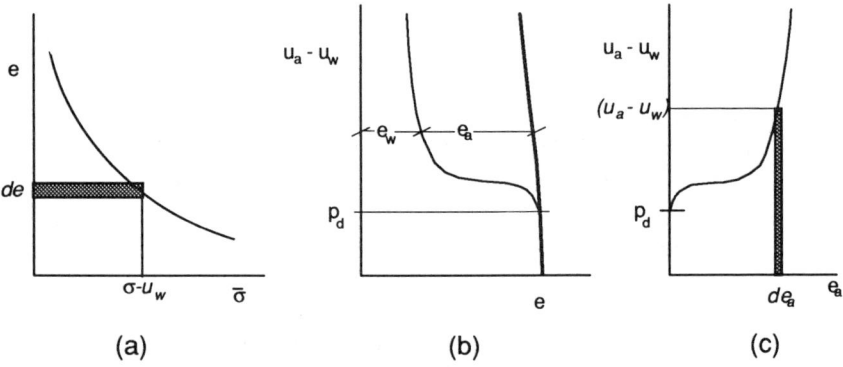

Figure 2 - Work Terms Defined in Equation 19

The same effect occurs if the air pressure is maintained at atmospheric pressure and the water pressure decreases due to drainage or evaporation. Increasing the total stress, increasing air pressure, and decreasing water pressure all cause the same one dimensional deformation as long as the soil is saturated. Therefore, as long as the capillary stress p_c remains below the displacement pressure, p_d, Eq. 19 can be written

$$\frac{dE}{V_s} = -(\sigma + (u_a - u_w)_{p_c < p_d})de \ . \tag{21}$$

This still describes effects on a standard e vs. $log \ \overline{\sigma}$ curve but also properly accounts for one dimensional deformations which occur when air pressure is increased or water pressure is decreased.

This relationship changes as soon as the displacement pressure is reached and air enters the soil. The air pressure, being a neutral stress, acts equally on the expanding interface as shown in Fig. 4. This forces the soil to dfoorm laterally as well as vertically since the capillary stress created by the interface is the same in both directions. In some soils and especially when the vertical total stress is zero or small compared to the capillary stress, the amount of lateral deformation may be very large

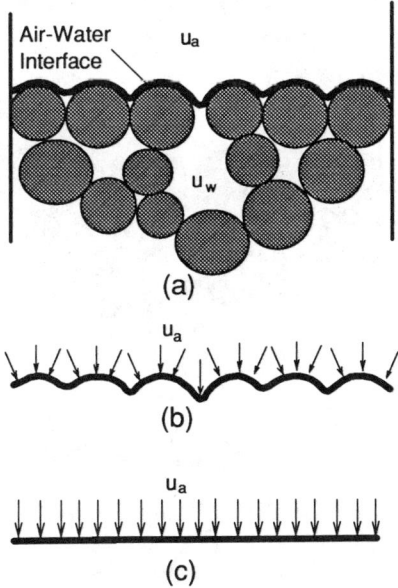

Figure 3 - Effect of Air-Water Interface In a Saturated Soil

because the normal stress perpendicular to the opening is now equal to the stress that
was acting on the surface. This manifests itself in the form of cracks in the soil. Thus,
soils will begin to crack at the displacement pressure. Cracking may proceed rapidly
after that change in state.

The amount of cracking will depend on the soil structure and fabric which is
dependent on many factors. These factors include the:
1) Ratio of the sand, silt and clay (i.e., the grain size distribution) in the soil,
2) Water content of the soil at deposition,
3) Time of settlement available before drying begins,
4) Segregation of the particles causing layering (the anisotropy),
5) Salt content of the soil,
6) Total stress acting on the soil,
7) Prior history of the deposit, and
8) Many other factors involved with sedimentation, erosion, desiccation, etc.

A fundamental question arises at this point. Is the void ratio defined in the
total volume including the crack space or is it defined just in the remaining clods? It is
typical to use the void ratio change as a measure of the vertical deformation (McNabb,
1960; Smiles and Rosenthal, 1968; Philip and Smiles, 1969; Edgar, Nelson and
McWhorter, 1989). This implies that void ratio change occurs perpendicular to the
soil surface. Subsequent discussion will be based on the total volume void ratio
change.

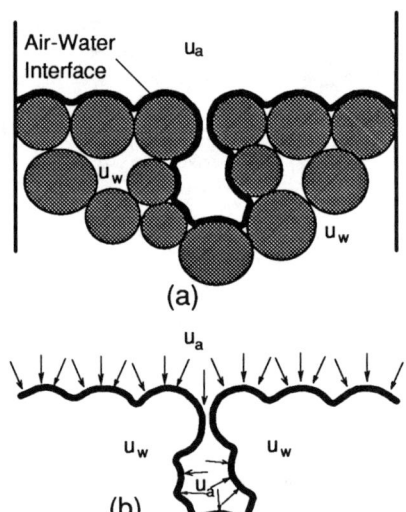

Figure 4 - Effect of Air-Water Interface on Unsaturated Soil

[An analysis of three dimensional finite strain (Edgar and Nelson, 1992) shows that the total volume void ratio must be defined by three components which describe the distribution of solid in the three material coordinate directions. In this way, the vertical deformation is a component of the total void ratio change.]

Constitutive Surface

Equation 19 suggests a constitutive surface such as that shown in Figure 5 in which the void ratio changes due to changes in effective stress and capillary stress. The vertical axis located at (0,0) is the void ratio. The axis to the left is the effective stress defined in Eq. 21. The straight line on this plane is the load line or the virgin consolidation curve. As a normally consolidated soil, the load line has been extensively investigated in the soil mechanics literature and the logarithmic relationship has been shown to be a good approximation. By extension, it also indicates the maximum past pressure and rebound curves must be able to fit within this constitutive surface. Because the capillary stress is less than the displacement pressure, the soil remains saturated while on the effective stress plane.

When the capillary stress p_c equals the displacement pressure p_d, the relationship changes and the soil begins to deform laterally as well as vertically. The ratio of the capillary stress to displacement pressure increases from a value of one, thus the logarithm increases from zero. Based on the author's visual observations of deforming soils in a pressure chamber, the logarithmic relationship appears to parallel

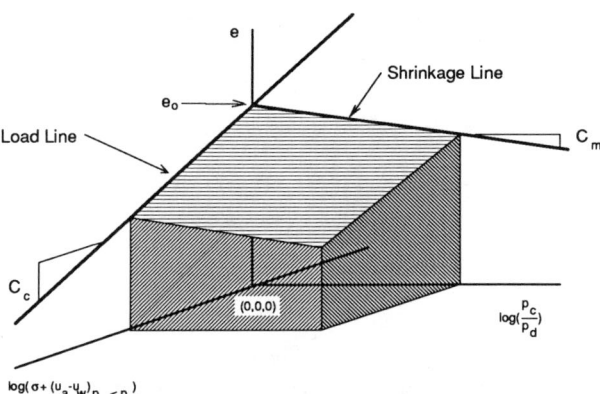

Figure 5 - Constitutive Surface Developed from Equation 19

the nature of the deformation, first changing rapidly and then more slowly as the capillary stress increases.

The surface shown in Fig. 5 is defined by the relationship,

$$e = e_o - C_c \cdot \log\left(\sigma + (u_a - u_w)_{p_c <= p_d}\right) - C_m \cdot \log\left(\frac{p_c}{p_d}\right)$$ (24)

Standard soil mechanics techniques can be used to account for overconsolidated conditions on the effective stress plane. Similar conditions can be applied on the capillary stress plane if that stress decreases. For example,

$$e = e_o - C_c \cdot \log\left(\sigma + (u_a - u_w)_{p_c <= p_d}\right)_{max} - C_r \cdot \log\left(\frac{\sigma + (u_a - u_w)_{p_c <= p_d}}{\left(\sigma + (u_a - u_w)_{p_c <= p_d}\right)_{max}}\right)$$
$$- C_m \cdot \log\left(\frac{p_{c_{max}}}{p_d}\right) - C_b \cdot \log\left(\frac{p_c}{p_{c_{max}}}\right)$$ (25)

where the terms with the subscript "max" represents the maximum past stress and C_r and C_b are the coefficients of the rebound curves on the effective stress and capillary stress axes, respectively.

The model also allows for the change in displacement pressure caused by application of the total stress. As the void ratio decreases along the load line, the value of displacement pressure increases because the smaller voids can sustain greater capillary stress differences.

Embedded in this relationship is another for the degree of saturation, as well. Figure 6 shows the capillary stress plane. The water void ratio, e_w is equal to

$$e_w = Se .$$ (26)

The ratio $\left(\frac{p_c}{p_d}\right)$ leads to a common relationship between capillary stress and the effective saturation S_e developed by Brooks and Corey (Corey, 1986),

$$S_e = \left(\frac{p_d}{p_c}\right)^\lambda = \frac{(S - S_r)}{(1 - S_r)}$$ (26)

where S_r is the residual saturation and λ is the pore size distribution index. There does not appear to be any information on the changes in residual saturation or pore size distribution index as a function of change in void ratio. Brooks and Corey have extended their relationship to estimate the hydraulic conductivity of partially saturated soil for water and air, as well.

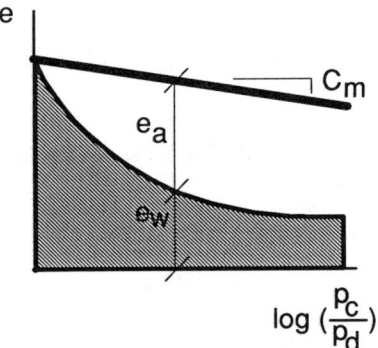

Figure 6 - Relationship of Water and Air Void Ratios on the Capillary Stress Plane

The constitutive surface defined in Eq. 24 has two limitations. At very high capillary stresses, the total volume approaches a constant value under a given total stress. Groenevelt and Bolt (1972) have shown this relationship as a function of water content and total stress as in Fig. 7. Equation 24 does not specifically have a limiting value. For most soils, very little deformation occurs above a capillary stress of 15 bars. This capillary stress could be used as an upper limit for the deformation calculation. A similar concern exists for the effective stress relationship. This is true of all consolidation relationships and is probably not a severe limitation.

Conclusion

A theoretical background has been developed for a deformation constitutive relationship based on physically defined macroscopic stress state variables. It accounts for the transition from saturated to unsaturated conditions. It is a simple and easy to implement relationship which can be easily extended for overconsolidation and cycling capillary stresses. Testing is currently being performed at the University of Wyoming to verify this relationship.

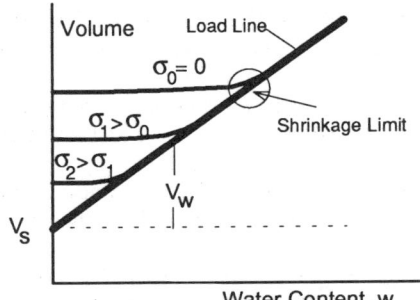

Figure 7 - Volume versus Water Content as a Function of Total Stress

Bibliography

Bishop, A.W. (1959) "The Principle of Effective Stress," Teknisk Ukeblad, Vol. 106, No. 39, 859-863.

Corey, A. T. (1986) Mechanics of Immiscible Fluids in Porous Media. Littleton, CO, Water Resources Publications,

Edgar, T.V. (1983) "Moisture Movement in Nonisothermal Deformable Media," Ph.D. Thesis, Department of Civil Engineering, Colorado State University, Ft. Collins, CO.

Edgar, T. V., Nelson, J. D. and McWhorter, D. B., (1989) "Nonisothermal Consolidation in Unsaturated Soil," Journal of Geotechnical Engineering, Vol. 115, No. 10, 1351-1372.

Edgar, T. V. and Nelson, J. D. (1992) "Flow Equations in Three Dimensional Finite Strain," Proceedings, 7th International Conference on Expansive Soils, ASCE, Dallas.

Fredlund, D. G. and Morgenstern, N. R. (1976) "Constitutive Relations for Volume Change in Unsaturated Soils," Canadian Geotechnical Journal, Vol. 13, No. 3, 261-276.

Fredlund, D. G. and Morgenstern, N. R. (1977) "Stress State Variables for Unsaturated Soils," Journal of Geotechnical Engineering Division, ASCE, Vol. 103, No. GT5, 447-466.

Groenevelt, P. H. and Bolt, G. H. (1972) "Water Retention in Soil," Soil Science, Vol. 113, No. 4, 238-245.

Haase, R. (1969) Thermodynamics of Irreversible Processes. Reading, Massachusetts, Addison-Wesley Publishing Co.

McNabb, A. (1960) "A Mathematical Treatment of One-Dimensional Soil Consolidation," Quarterly of Applied Mathematics, Vol. 17, No. 4, 337-347.

Philip, J. R. and Smiles, D. E. (1969) "Kinetics of Sorption and Volume Change in Three-Component Systems," Australian Journal of Soil Research, Vol. 7, No. 1, 1-19.

Smiles, D. E. and Rosenthal, M. J. (1968) "The Movement of Water in Swelling Materials," Australian Journal of Soil Research, Vol. 6, No. 4, 237-248.

Terzaghi, K. (1925) Erdbaumechanic. Vienna: Franz Deutcke.

FLOW PROPERTIES OF UNSATURATED SOILS
UNDER CONTROLLED SUCTION

J.F.T.Jucá [1]

ABSTRACT

Laboratory tests were performed for the hydraulic characterization of three compacted soils under high suction. Moisture changes were studied considering some factors affecting the time of moisture equilibrium. Suction-moisture content tests were conducted in order to obtain data on soil water retention, suction-moisture capacity, and suction-moisture content-void ratio relationships. The hydraulic conductivity of these soils decreased about 4 to 5 orders of magnitude, relative to the saturated hydraulic conductivity, when the suction values increased from 0 to 400kPa.

INTRODUCTION

The performance of engineering structures on unstable soils, and the process of groundwater contamination are largely dependent on moisture movement in unsaturated soils. The measurement of the flow properties of such soils is related to soil water retention and hydraulic conductivity.

In general, field methods used to determine the unsaturated soil hydraulic properties are usually time-consuming, expensive, and subject to simplifying assumptions. Laboratory methods suffer from similar problems, but allow measurement of the hydraulic conductivity of the soil following both wetting and drying paths under controlled suction and vertical stress. An alternative for a simple characterization of the hydraulic properties of unsaturated soils is the calculation of the conductivity and the differential moisture capacity from suction-water content data, which is readily obtained.

The apparatus and techniques used to measure the unsaturated soil properties are

[1] *Professor of Civil Engineering Department, Federal University of Pernambuco, Centro de Tecnologia, Cidade Universitária, 50740-530 - Recife - PE, Brazil.*

dependent on the soil suction range. At low suctions, the soil variables that could be measured allow a full characterization of the soil behavior. At high suctions (above 1.5MPa), however, it is more difficult to determine the soil stress state. The soil suction values can not be directly measured, and the tests are conducted under controlled suction. In this case, the soil moisture changes and the time of moisture equilibrium are relevant parameters for monitoring the tests.

The purpose of this paper is to present a method for laboratory hydraulic characterization of unsaturated soils under high suction values. Moisture changes, suction-moisture content relationships, and hydraulic conductivity values are presented, using three types of compacted soils covering a wide range of materials. Previous works has shown the strength and deformation behavior of these soils (Escario and Sáez, 1986; Escario and Jucá, 1989; Jucá and Escario, 1991).

EXPERIMENTAL PROCEDURES

Laboratory tests were conducted for the characterization of three soils, which cover a wide range of materials. The first is "Madrid grey clay", a high plasticity clay (described in this paper as Gray clay). The second is "Guadalix red clay", a silty clay with low plasticity (Red clay), and the last is a sand with a small amount of clay (termed Clayey sand). Samples were prepared at predetermined moisture content and density using a static compaction method. The properties of these soils and the initial conditions of the samples are summarized in Table 1.

Table 1. Properties of the soils and initial conditions of samples

Soil	Gray clay	Red clay	Clayey sand
Texture (%)			
Sand	1	17	87
Silt	27	48	7
Clay	72	35	6
Atterberg limits			
LL	71	33	28
PI	35	13.6	8
Specific gravity, G_s	2.71	2.66	2.64
Standard Proctor			
γ_{max}(kN/m³)	13.4	18.0	19.1
w_{opt}(%)	33.7	17.0	11.5
Compacted samples			
γ_d(kN/m³)	13.4	18.0	19.1
w(%)	29.0	13.6	9.2
Initial suction (MPa)	0.80	0.28	0.07
Void ratio, e_o	1.03	0.48	0.38
Initial degree of saturation, S (%)	76	75	64

Tests were carried out in different devices developed for running tests under controlled suction, based on the axis translation technique (Hilf, 1956).

Standard pressure cells (ASTM D3152-72) were used for the determination of the suction-moisture content relationship, and the time of moisture equilibrium. Samples were compacted into a 42mm diameter, 20mm high mold. Twenty seven tests were carried out for suction-moisture characterization of the three soils. The suction ranges were different for each soil. For the Gray clay, the maximum value reached was 12MPa, while for the Red silty clay and Clayey sand, the values were 8MPa and 4MPa, respectively.

During a majority of the tests, moisture changes and the time of moisture equilibrium were observed. Time dependence arises from the impedance of the system soil/contact/porous plate or membrane. In order to study the effects of the contact between the samples and the membrane, half of the tests were conducted with a calibrated spring placed on top of the sample. The spring, holding the sample against the membrane, was selected to apply a pressure of about 10kPa.

Hydraulic conductivity tests were performed using an oedometer cell developed by Escario (Escario and Sáez, 1973). The tests were running under controlled vertical stress, and suction. In this test, the sample (70mm diameter and 20mm height) is placed on a saturated porous plate in the cell, under a given vertical stress. Increments of air pressure (or suction) were applied generating excess of pore water pressure in the sample, while the water in the plate is maintained at atmospheric pressure. The suction values were increased from 0 to 450kPa, in steps of 50kPa. The sample is allowed to drain out the excess water. The volume of outflow as a function of time is measured during all steps of the moisture equilibrium, while monitoring the vertical strain of the sample.

The hydraulic conductivity values were obtained from outflow data, using the method published by Kunze and Kirkham (1962) which takes non-negligible plate impedance into account.

Both pressure and oedometer cells present problems with minuscule quantities of outflow, impedance effects from porous plates or membranes, and diffusion of gas into the measuring system. These problems were considered during the experiments. Full details of the testing equipment and techniques are given by Jucá (1990).

MOISTURE CHANGES

The moisture changes under suction and the moisture equilibrium process were studied considering the effects of contact between the sample and the membrane on the time of equilibrium. The soil moisture changes with time are presented in Figure 1, as the *Percentage of Moisture Equilibrium, U_w*. Its value is the same as the soil moisture concentration, where w, w_o and w_f are the sample moisture content at times t, initial, and final, respectively. The final moisture content values are the same as the moisture contents at equilibrium. These values were used to determine the suction-moisture content relationships.

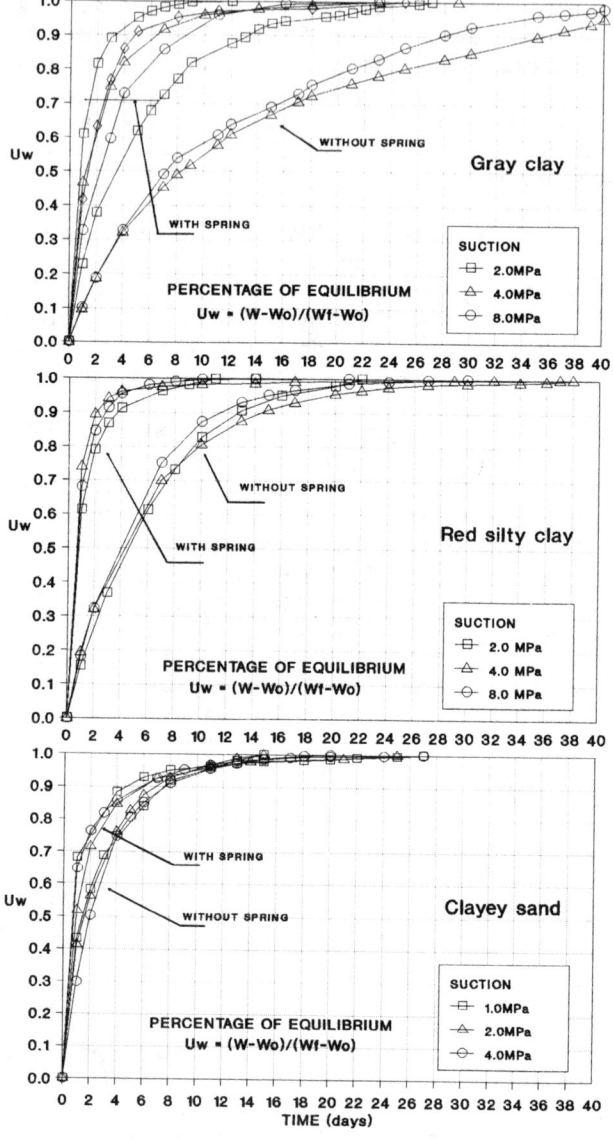

Fig.1. Percentage of Moisture Equilibrium, U_w, versus Time.

The percentage of moisture equilibrium was determined for the three soils tested under different suctions with and without the spring. Figure 1 shows the results of the tests which followed a drying path, although another series of tests was performed following a wetting path. As may be seen, for the same percentage, the time of moisture equilibrium is larger for the Gray clay than for the Red silty clay, in accordance with their soil properties. On the another hand, the Clayey sand had a longer time of equilibrium than the Red silty clay. This unexpected result can be partly explained by the poor contact of the sandy soils with the porous plate or membrane and to its small hydraulic conductivity at a low degree of saturation.

It can also be seen in Figure 1, that the moisture equilibrium is reached much faster when a slight pressure is applied to the sample. In this type of test, a condition of good contact is necessary to quantify the time of moisture equilibrium. These findings are similar to the results presented in the literature, including tests performed under low suction values.

In general, the time of moisture equilibrium is mainly affected by the moisture path followed, the suction gradient, and the impedance of the drainage system. When the impedance is reduced, the tests show a relatively short time of equilibrium, even for high suction values. For the same impedance, the moisture changes are progressively slower as the suction increases.

SUCTION - MOISTURE CONTENT RELATIONSHIPS

The relationship between suction and moisture content has been largely used for the characterization of the soil hydraulic properties, although recently this function has been used to link the soil suction data with the soil at a critical state (Brady, 1988; Tool, 1990).

The suction tests described in this paper were performed in order to define suction-water content-void ratio relationships. They were also used to verify the effect of the small pressure applied by the spring, on the final soil moisture content.

Figure 2 gives the suction-water content relationships of the three soils tested. It shows the relative position and the typical shape of the characteristic curve for the different materials. This behavior depends on the structure of the soil, the texture, and the specific surface area of the soil material. In clayey soils, with a large specific surface area, more water is adsorbed than in silty or sandy soils. The shape of the curves is related to the pore-size distribution, with a gentle shape for well-graded soils and a more abrupt change in slope when the pore-size is more uniform. It can also be seen from Fig.2 that there is a good agreement between the points defined by "with-spring" tests and the "without-spring" tests. This fact suggests that the effect of the small pressure on the final soil moisture content is not large.

The slopes of the characteristic curves were used to obtain the differential water capacity of the soils, which is termed the *Suction-Moisture Capacity, C.* It is

defined as the amount of moisture gain or loss per unit change of soil suction on
the logarithm or pF scale. The value of C depends upon the soil properties(texture,
consistency limits), the moisture range, and hysteretic effects. For practical
purposes, its value can be taken as a constant (for a moderately small suction
range) defining wetting or drying behavior (Fig.3). In general, the value of C
increases with the clayey fraction and the plasticity index. It is equivalent to the
soil moisture capacity of retention.

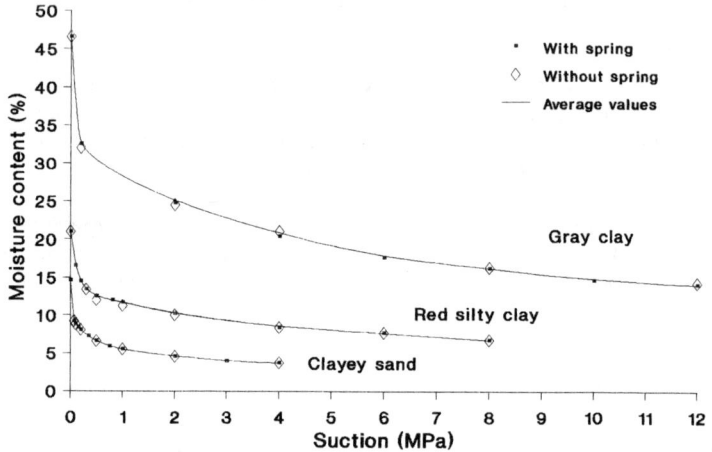

Fig.2. Suction-moisture content relationships.

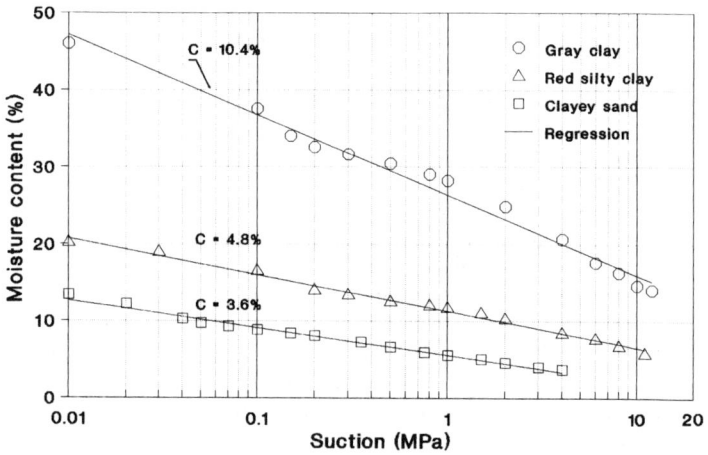

Fig.3. Suction-Moisture Capacity, C.

An attempt to determine which soil properties were most strongly linked with the suction-moisture capacity, C, was made. For silty and clayey soils, linear relationships were obtained by Jucá (1990), using hygroscopic moisture content, liquid limit, and plastic index. In this case, the amount of experimental data available was limited to define a general trend.

These data were combined with compacted soils data from Brazil and Spain in order to obtain a relationship between the suction-moisture capacity and the liquid limit. The line of best fit is presented in Figure 4.

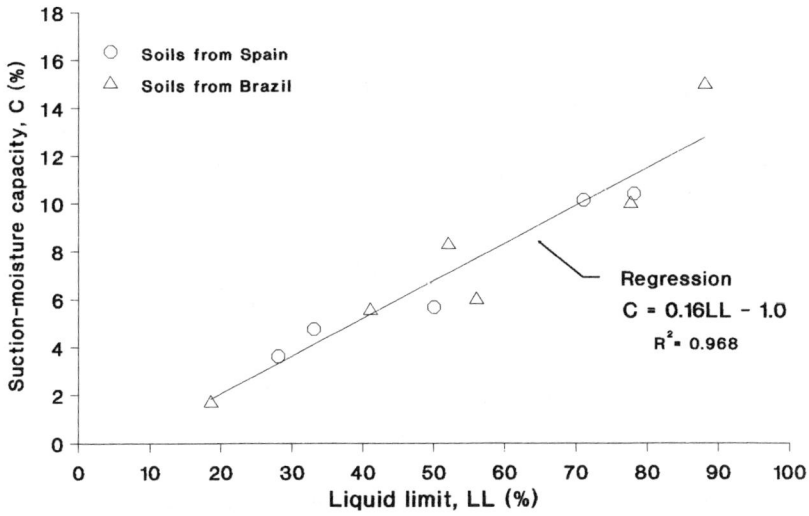

Fig.4. Suction-moisture capacity and liquid limit.

Other suction tests were performed following drying and wetting paths in order to observe the hysteretic behavior of the three soils. The samples were initially brought to equilibrium at their initial suction. The suction was then increased and the soil followed a drying path. The wetting path was obtained by reducing the suction, defining a primary loop from each soil (Fig.5). Secondary drying and wetting paths were followed, but not shown on Fig.5.

It can be seen that the suction-moisture content paths for the soils are very similar, following roughly parallel paths for the Clayey sand and Red silty clay. The Gray clay showed a more hysteretic effect than the other two soils. This behavior can be explained by their clay properties, including shrinking and swelling under moisture changes. In general, hysteresis is appreciable when the soil is subjected to alternate wetting and drying.

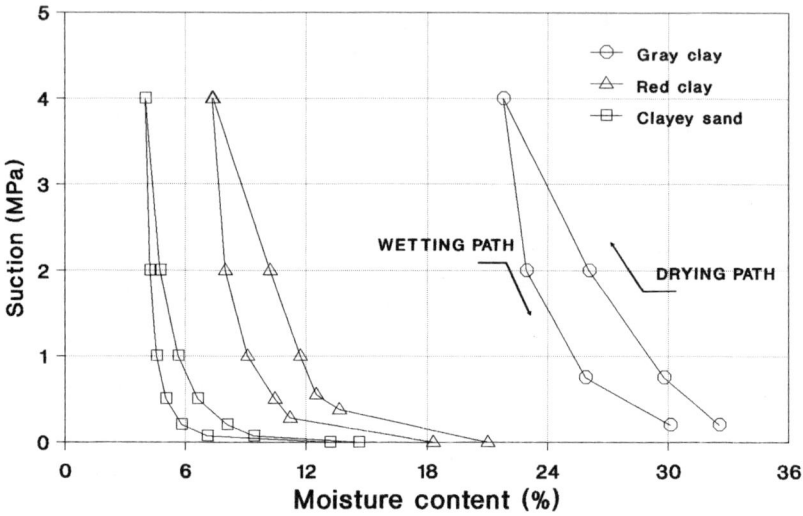

Fig.5. Hysteretic behavior of the soils.

SUCTION - MOISTURE CONTENT - VOID RATIO RELATIONSHIP

A series of shrinkage tests was carried out on the Gray clay. Samples were prepared with 42mm diameter and 20mm height at the same moisture content and statically compacted at the same densities. The samples were then allowed to dry on the pressure cells at different suctions. When moisture equilibrium was obtained, the samples were weighed and volume determined. The results of the tests are shown in Figure 6, where the decrease of the void ratio with an increase in the suction can be seen.

The results of the tests show a typical suction-void ratio relationship for a clayey soil. The void ratio values are varied from 1.03 to 0.86 in the suction range between 0.8MPa and 12MPa following a drying path.

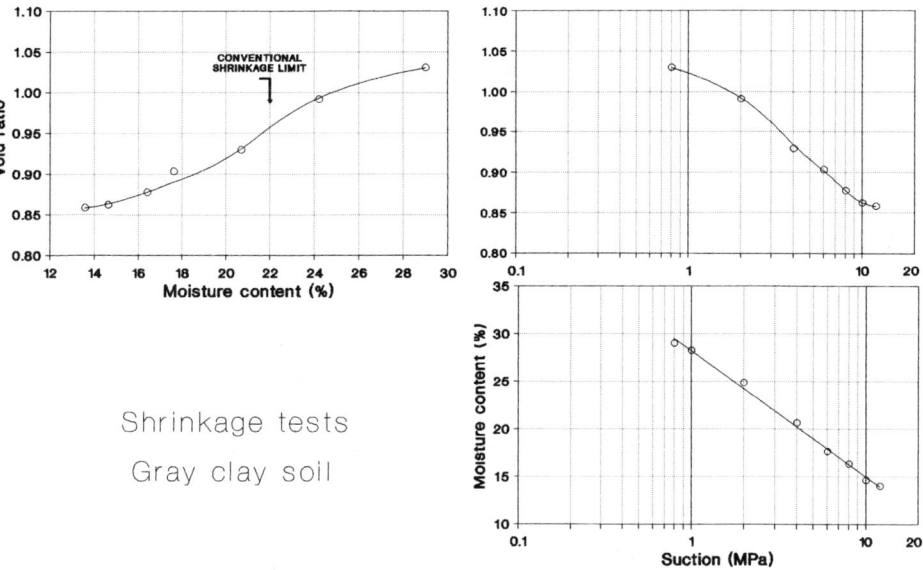

Shrinkage tests

Gray clay soil

Fig.6. Suction - moisture content - void ratio relationship.

HYDRAULIC CONDUCTIVITY

The hydraulic conductivity values were determined from the incremental outflow method using an oedometer cell under controlled vertical stress, and suction values of up to 450kPa. At higher suction values, the data were obtained from membrane pressure cells, using a method which takes any non-negligible plate impedance into account.

Hydraulic conductivity tests were performed on the Clayey sand in order to verify its moisture changes under suction. The vertical stress applied was 10kPa in order to compare results with those from the suction tests performed with the same stress. The suction values were increased from 0 to 450kPa, in steps of 50kPa.

Figure 7 gives the results of these tests. As may be seen, the hydraulic conductivity values are strongly affected by soil suction. The higher values for the 0 - 70kPa range suggests that there may be a very sharp transition in this range. These values were compared with the hydraulic conductivity obtained in saturated conditions giving a relative hydraulic conductivity Kuns/Ksat of about 10^{-5}.

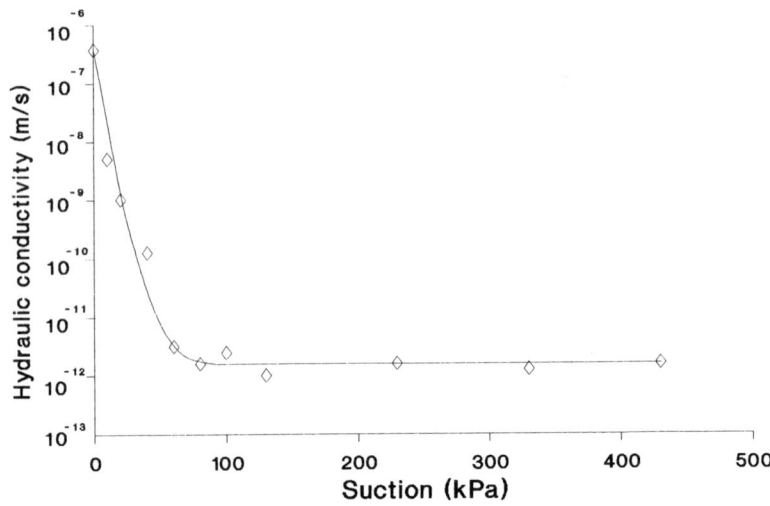

Fig.7. Hydraulic conductivity of the Clayey sand.

An alternative to direct measurements is the calculation of the hydraulic conductivity from soil water retention data. The predictive methods for hydraulic conductivity require accurate descriptions of the soil water retention curve. In this way, the suction-moisture content relationship was used on Mualem's model, combined with the Van Genuchten's equation (Mualem, 1976; Van Genuchten and Nielsen, 1985; Nielsen et al, 1986).

The results are given in Figure 8 as the hydraulic conductivity relative to the saturated permeability. The saturated permeability values were obtained in standard triaxial cells for the three soils, using back pressure to achieve saturation of samples. All the results in Table 2 fall in a fairly narrow range for hydraulic conductivity values.

Table 2. Saturated permeability

Soil	Saturated permeabilities (m/s)
Gray clay	2.8×10^{-9}
Red clay	1.0×10^{-8}
Clayey sand	5.7×10^{-7}

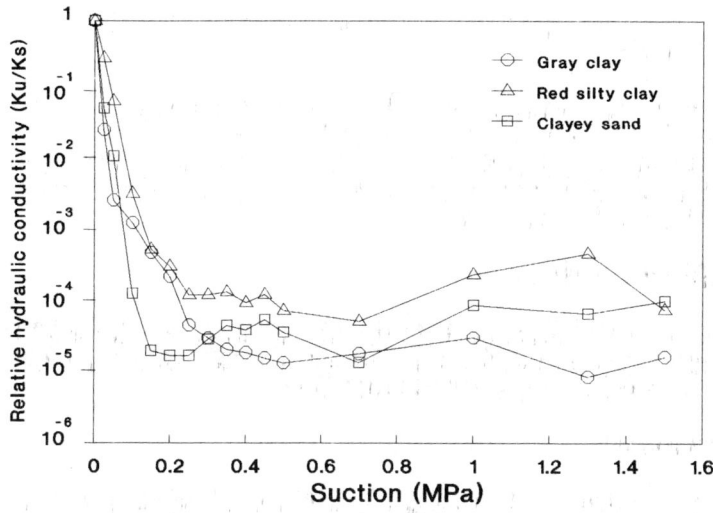

Fig.8. Relative hydraulic conductivity.

CONCLUSIONS

- The time of moisture equilibrium in the pressure membrane cell is more shorter if a good contact with the membrane is assured by applying a small vertical pressure on the soil sample.

- Hydraulic conductivity tests, performed in oedometer cells under controlled vertical stress and suction, allow measurements of the water outflow or inflow coupled with volume changes of the soil.

- The time of moisture equilibrium is affected by the moisture path followed, the suction gradient and the impedance of the drainage system. When the impedance is reduced, the moisture change results show a relatively short time of equilibrium, even for high suction values.

- Under unsaturated conditions, the soil hydraulic properties could not be directly related to saturated values. It was found that the time of moisture equilibrium in sandy soils is longer than in clayey soils. At a low moisture content, the water in the granular soil does not make a good contact with drainage systems in order to facilitate continuous flow. In this case, a major part of the moisture movements occurs by vapour transfer, but under low velocity.

- A linear relationship between the suction-moisture capacity and liquid limit could be established.

- The hydraulic conductivity of the three soils decreased about 4 or 5 orders of magnitude, relative to the saturated hydraulic conductivity, when the suction values increased from 0 to 400kPa. In this suction range, the Clayey sand exhibited a degree of saturation of 55 percent, while for the Gray clay and the Red silty clay, this value was about 70 percent.

ACKNOWLEDGMENT

A majority of the tests and research described in this paper were carried out in the Laboratorio de Geotecnia (CEDEX), Madrid. The contribution made by Dr.Ventura Escario is gratefully acknowledged.

REFERENCES

Brady, K.C.(1988)."Soil suction and the critical state".Geotechnique38,1,117-120.

Escario, V. and Jucá, J.F.T. (1989). "Strength and deformation of partly saturated soils". Proc. of the 12th Int. Conf.on Soil Mech. Found. Eng., Rio de Janeiro, 1, 43-46.

Escario, V. and Sáez, J. (1973). "Measurement of the properties of swelling and collapsing soils under controlled suction". 3rd Int. Conf Exp. Soils, Haifa, 1, 195-200.

Escario, V. and Sáez, J. (1986). "The shear strength of partly saturated soils". Geotechnique, Vol. 36, 3, 453-456.

Hilf, J.W. (1956). An investigation of pore-water pressure in compacted cohesive soils. Technical Memorandum N° 654, Bureau of Reclamation, Colorado.

Jucá, J.F.T. (1990). Behavior of unsaturated soils under controlled suction (in spanish). Doctor Thesis, Politechnical University of Madrid, Spain.

Jucá, J.F.T. and Escario, V. (1991). Variation of the modulus of deformation of soils with suction. X European Conference on Soil Mechanics and Foundation Engineering, Firenze, Vol 1, 121-124.

Kunze, J.K. and Kirkham, D. (1962)."Simplified Accounting for membrane impedance in capillary conductivity determinations". Proc. of Soil Science. Soc. Am.,29(5),421-426

Mualem, Y.A. (1976). A new model for predicting the hydraulic conductivity of unsaturated media. Water Resour. Res., Vol. 12(3), 513-522.

Nielsen, D.R.; Van Genuchten, M.Th. y Biggar, J.W. (1986). "Water flow and solute transport processes in the unsaturated zone". Water Res.Res.22(9),89S-108S.

Toll, D.G. (1990). "A framework for unsaturated soil behavior". Geotechnique 40, N° 1, pp. 31

Van Genuchten; M.Th. and Nielsen, D.R. (1985). "On describing and predicting the hydraulic properties of unsaturated soils". Ann. Geophys., 3, 615-628.

Permeability Determination For Unsaturated Soils

A.Naser Abu-Hejleh [1], Dobroslav Znidarčić [2], and Tissa H. Illangasekare [3]

ABSTRACT

A reliable approach for the determination of the soil hydraulic conductivity function relies on the parameter estimation solution approach in which heterogeneous state of saturation within the sample is accounted for in the analysis of the test data. Accurate test data needed for the analysis can be obtained from a transient flow experiment in which water is drained from soil samples under controlled flow rates using a flow pump. The feasibility of this approach is investigated with a parameter estimation algorithm and a non-linear finite element program which numerically simulates the experimental conditions for any set of constitutive parameters. In the analysis, the soil hydraulic conductivity function is assumed to be represented by the van Genuchten's expression which involves two constitutive parameters, a_2 and n_2, that are assumed to be different from parameters a_1 and n_1 characterizing the water retention function.

INTRODUCTION

Emphasis on the flow processes in soils for the prediction of the movement of contaminants requires a detailed knowledge of the hydraulic characteristics of soils. These include not only the saturated hydraulic conductivity but the relative permeability in the unsaturated regime as well. Unsaturated conditions are routinely found in field situations when flow through the compacted clay liners is considered or when the movement of contaminants through the vadose zone above the groundwater table has to be evaluated. The techniques used for the direct measurement of the unsaturated hydraulic conductivity are often cumbersome and require elaborate instrumentation for measuring pore pressure changes and

---

[1]Research Associate, Department of Civil, Environmental and Architectural Engineering, University of Colorado, Boulder, CO 80309.

[2]Associate Professor, Department of Civil, Environmental and Architectural Engineering, University of Colorado, Boulder, CO 80309.

[3]Professor, Department of Civil, Environmental and Architectural Engineering, University of Colorado, Boulder, CO 80309.

fluid movement within specimens during testing. The problem complicating the direct measurement of the unsaturated hydraulic conductivity is the difficulty in maintaining a homogeneous degree of saturation within the sample during testing. Namely, in order to measure permeability the fluid must flow through the specimen and in the unsaturated condition this leads to the continuous change of degree of saturation in most cases. Some of the laboratory techniques for the determination of the unsaturated hydraulic conductivities in soils are described by Olson and Daniel (1981).

An approach to the measurement of the hydraulic conductivity of unsaturated soils is not to try to create a homogeneous degree of saturation within the sample but to recognize in the test analysis that the degree of saturation changes within the sample throughout the test. This is the so called inverse problem solution approach. In this approach the test is considered to be a boundary value problem for which the form of the governing equation and boundary conditions are known while the constitutive relations are unknown. The measurements in the experiment define the solution to the problem. Since only the constitutive properties are unknown, they can be determined through the inverse problem solution process in which the measurements are matched in the analysis. Such an approach has been successfully used for the determination of the consolidation and desiccation characteristics of soft soils, for which the test conditions also create the heterogeneous state of permeability within the sample (Znidarcic and Liu, 1989; Abu-Hejleh, 1993). Kool et al. (1985) have applied the same approach to the analysis of a draining sand column experiment in order to determine the constitutive properties for unsaturated sands. Their analysis demonstrated the need for precise measurements, confirming the fact that in most cases the inverse problems are mathematically ill posed, meaning that the small variation of the input data produces large variations of the results.

This paper presents an analysis for the determination of unsaturated soil hydraulic conductivity using the inverse problem solution approach for the controlled outflow test. The laboratory technique uses a flow pump for a precise control of outflow rates from the sample, enabling accurate determination of the change of the average water content of the sample with time. A precision differential transducer is used to measure the suction development at at the bottom of the sample. The technique was used by Manna et al. (1993) to determine the suction-saturation relationship, and it is the purpose of this paper to show how the test results of this technique can be used to estimate the hydraulic conductivity relationship as well. Compared to the method of Kool et al. (1985) the advantage of the flow pump experiment is in the much higher accuracy of the input data for the analysis. In addition, the possibility of obtaining a non unique solution is reduced in the analysis by determining the suction- saturation relationship in a separate test.

As the first step in the evaluation of the proposed methodology an actual test is simulated in the numerical analysis and the results are used as input data for the parameter estimation algorithm. By taking this approach a rational evaluation of the method is possible without being affected by the question of measurement accuracy that is inevitable in an actual test. The final verification of the method, using the actual test results, we leave for a later paper. The aim of this paper is to present the concept and demonstrate the feasibility of the approach.

THEORY

The governing equation for one-dimensional water flow for unsaturated soil is given by Richards (1931) as:

$$C(h)\frac{\partial h}{\partial t} = \frac{\partial}{\partial z}\left[K(h)(\frac{\partial h}{\partial z} - 1)\right] \tag{1}$$

where z refers to the vertical coordinate taken positive downward; h is the pressure head given as $h = u/\gamma_w$ in which γ_w is the unit weight of water and u is the pore water pressure; $C = \frac{d\theta}{dh}$ is the soil water capacity in which θ is the volumetric water content; and K is the hydraulic conductivity. The volumetric water content is expressed as $\theta = n.S$, where n is the porosity and S is the degree of saturation. For an essentially incompressible soil, such as sand considered here, the change in θ is proportional to the change in degree of saturation since porosity remains constant. The solution of Eq.1 requires the initial and boundary conditions and the constitutive relations, namely the water retention and the hydraulic conductivity functions, to be described.

The initial conditions and boundary conditions in the controlled outflow test are expressed mathematically as:

$$h = z \qquad t = 0, \qquad 0 \leq z \leq L \tag{2}$$

$$\frac{\partial h}{\partial z} = 1 \qquad t > 0, \qquad z = 0 \tag{3}$$

$$K(-\frac{\partial h}{\partial z} + 1) = v \qquad t > 0, \qquad z = L \tag{4}$$

where L is the height of the sample in which $z = 0$ is taken at the top and $z = L$ is taken at the bottom. The initial conditions described in Eq. 2 correspond to zero Darcian water velocity across the sample for which the pore water pressure is hydrostatic. The boundary conditions expressed in Eqs. 3 and 4 are zero Darcian water velocity at the top of the sample and the applied downward Darcian water velocity, v, at the bottom of the sample, respectively. These boundary conditions reflect properly the physical conditions in the test in which the water is removed at a constant flow rate from the bottom while only air (no water) is allowed to enter at the top.

The water retention and the hydraulic conductivity functions are assumed to be represented by van Genuchten's expression (van Genuchten, 1980) but with the constitutive parameters for the hydraulic conductivity function different from those controlling the water retention function. The water retention function is given as:

$$\theta(h) = (\theta_s - \theta_r)\frac{1}{[1 + |a_1 h|^{n_1}]^{m_1}} + \theta_r, \qquad h < 0$$

$$\theta = \theta_s, \qquad h \geq 0 \tag{5}$$

where θ_s is the saturated volumetric water content; θ_r is the residual volumetric water content; and a_1, m_1, and n_1 are empirical parameters such that $m_1 = 1 - 1/n_1$. The hydraulic conductivity function is given as:

$$K(h) = K_s \frac{\{1 - |a_2 h|^{n_2 - 1} [1 + |a_2 h|^{n_2}]^{-m_2}\}^2}{[1 + |a_2 h|^{n_2}]^{m_2/2}}, \qquad h < 0$$

$$K(h) = K_s, \qquad h \geq 0 \tag{6}$$

where K_s is the saturated hydraulic conductivity; a_2, m_2, and n_2 are empirical parameters such that $m_2 = 1 - 1/n_2$. Considering Eq. 5, the expression for $C(h)$ is obtained as:

$$C(h) = (\theta_s - \theta_r)m_1 n_1 a_1 \frac{|a_1 h|^{n_1 - 1}}{[1 + |a_1 h|^{n_1}]^{m_1 + 1}}, \qquad h < 0$$

$$C(h) = 0, \qquad h \geq 0 \tag{7}$$

The finite element technique was used to solve Eq. 1, taking into consideration the form of constitutive relations expressed in Eqs. 5 and 6. The element equations, which arise from the finite element discretization, are assembled into a global system of non-linear algebraic equations that are solved using an iterative scheme. The value of $C(h)$ is set equal to 1×10^{-6} instead of 0, whenever $h \geq 0$, for the ease of the numerical solution. The algorithm of the numerical solution was coded in a computer program named SEEP.

NUMERICAL EXPERIMENTS

The sample used in the numerical experiments has an initial thickness of 5.0 cm, and the soil characteristics correspond to a sand designated as #125, where 125 is the sieve opening in μm through which the sand had been sieved. The parameters of the water retention function for this sand were determined experimentally using the flow pump technique by Manna et al. (1993) as: $a_1 = 1.78$, $n_1 = 6.5$, $\theta_s = 0.41$, and $\theta_r = 0.09$, and the corresponding water retention function is shown in Figure 1 where the suction was calculated as $-\gamma_w h$. The

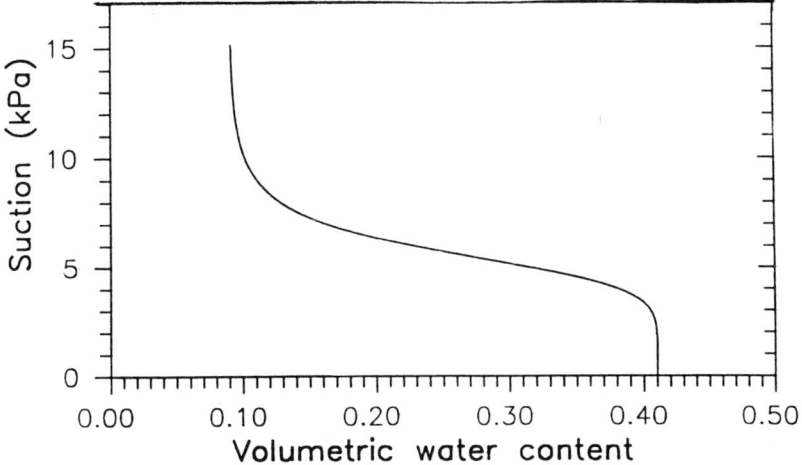

Figure 1: Water Retention Function for the Soil Used in the Numerical Tests.

saturated permeability for this sand, K_s, is 0.0684 m/hour. Three different hydraulic conductivity functions were assumed for this soil and their a_2 and n_2 values are: (i) 1.78 and 6.5, (ii) 2.0 and 7.0, and (iii) 1.5 and 6.0; these functions are shown in Figure 2. The soil sample was subjected to three values of bottom Darcian water velocities, v, given as (1) a low value of $\frac{K_s}{400}$, (2) a high value of $\frac{K_s}{10}$, and (3) a very high value of $\frac{K_s}{4}$. Program SEEP was used to simulate these tests and to predict their results. The number of nodes used to discretize the soil sample in the numerical simulation was 81 and the first time step was chosen as 1×10^{-6} hour. An adaptive technique, that assures the accuracy of the numerical solution, was used to update the subsequent time steps. It was concluded after performing several numerical simulations that discretizing the soil sample to number of nodes larger than 81 leads to insignificant change in the output results. The numerical simulation was terminated once the maximum calculated node suction at any iteration exceeded 15 kPa.

Samples of the numerical test results in term of the bottom suction with time under the low and the high values of v are shown in Figures 3 and 4. These figures show clearly the effect of the permeability function on the obtained progress of the bottom suction with time and indicate that this effect becomes more significant as the applied Darcian velocity increases and as the permeability of the soil decreases, and vice versa. The distribution of the suction within the soil sample at various times under the low and high values of v are shown in Figures 5 and 6. Figure 5 shows that low value of v induces homogeneous states of suction within

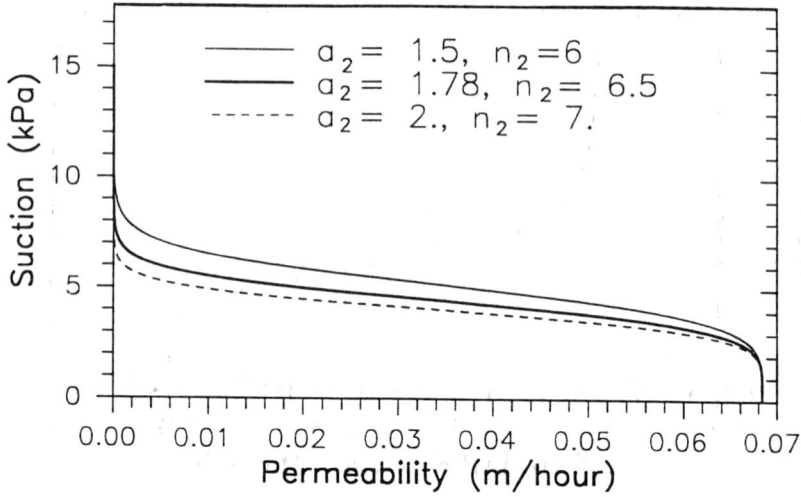

Figure 2: Permeability Functions for the Soil Used in the Numerical Tests.

the sample up to the final stage of the test, indicating that the permeability has insignificant effect on the obtained response. Figure 6 shows that high value of v induces heterogeneous states of suction near the bottom portion of the sample, indicating that the permeability has significant effect on the obtained response. In other words, the low values of v induce small gradient so that the suction distribution is identical to the static conditions up to the final stages of the test. For the high values of v, the gradient is high so that the suction distribution is controlled by the permeability characteristics of the sample. Hence, the proposed technique in this paper can lead to the reliable permeability characteristics if an appropriate v is chosen in the test. A low Darcian velocity that induces homogeneous degree of saturation within the sample will lead to a non unique solution, i.e, several combinations of the hydraulic conductivity parameters will match the test results. On the other hand, a very high value of v will terminate the test quickly leading to a potentially unreliable measurements.

In order to determine the suitable value of v for a given soil that assures homogeneous or heterogeneous states of suction within the sample, the results of the outflow controlled test should be drawn in term of the bottom suction versus the average volumetric water content for different values of v, as shown in Figure 7. As long as the obtained response coincides with the response obtained under the lowest value of v, homogeneous states of suction exist within the sample. Once the response starts to deviate from the response obtained under the lowest value of v, heterogeneous states of suction exist within the sample. The homogeneous

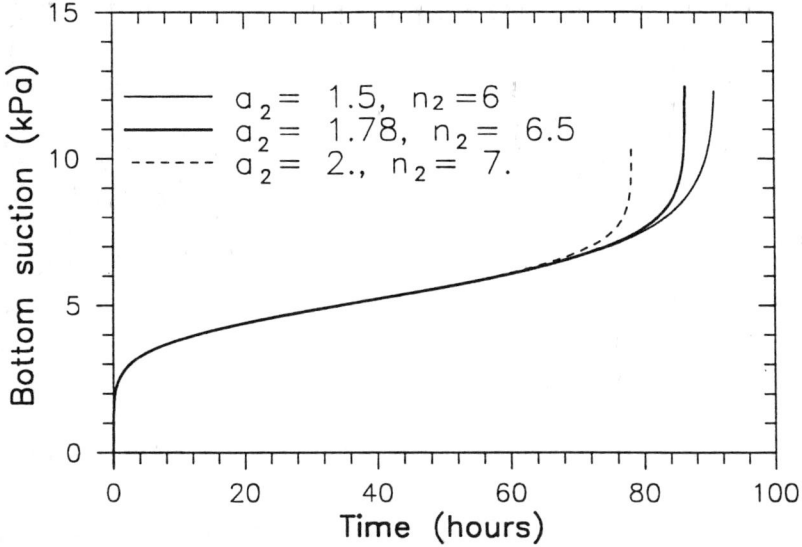

Figure 3: The Test Results When the Flow Rate is Low.

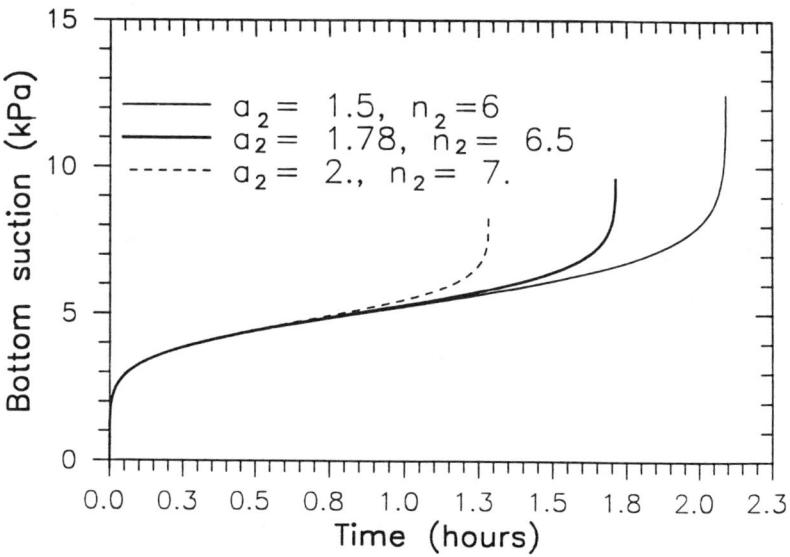

Figure 4: The Test Results When the Flow Rate is High.

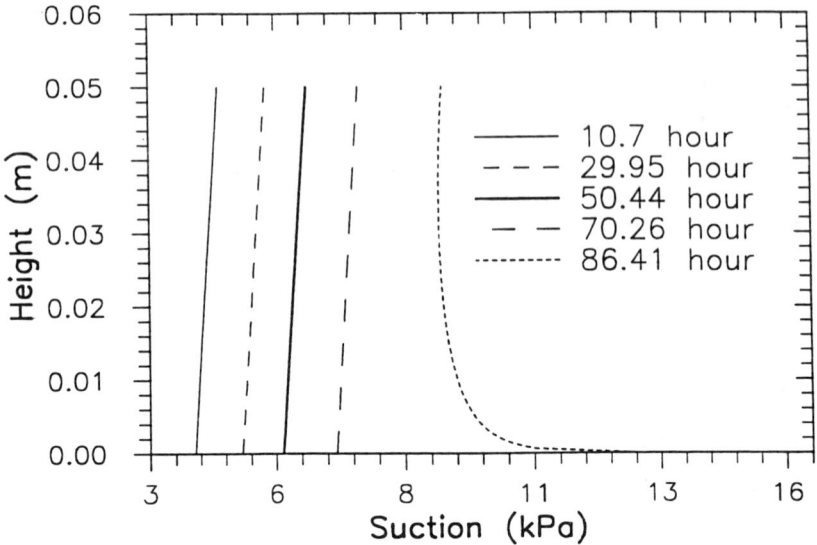

Figure 5: Distribution of Suction within the Sample When the Flow Rate is Low,
$a_2 = 1.78$ and $n2 = 6.5$.

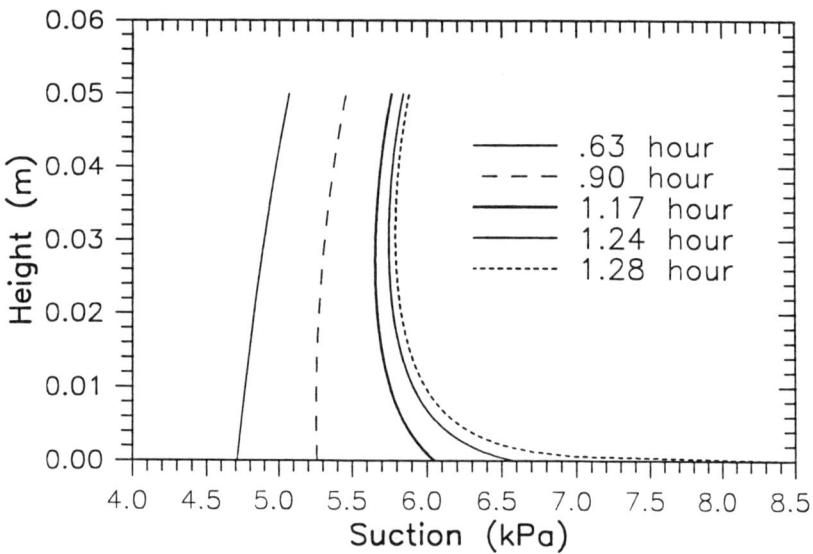

Figure 6: Distribution of Suction within the Sample When the Flow Rate is High,
$a_2 = 2$ and $n_2 = 7$.

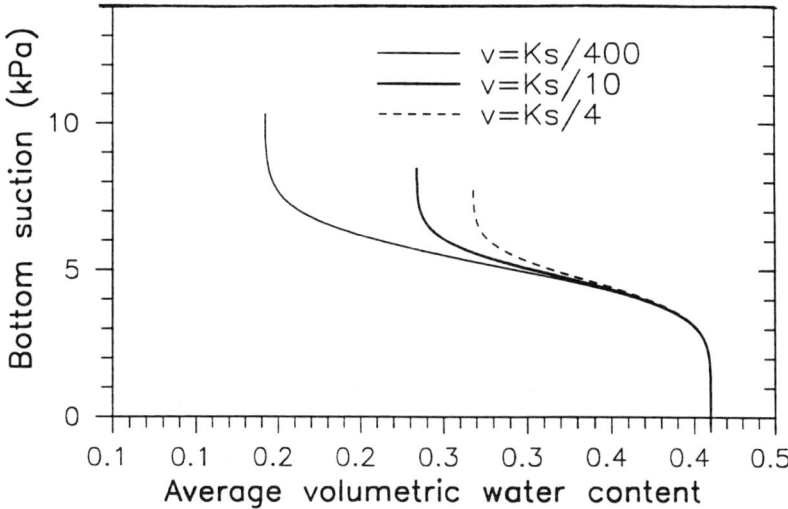

Figure 7: The Test Results Under Different Flow Rates, $a_2 = 2$ and $n_2 = 7$

states of suction, obtained under the low value of v, can be used to evaluate the retention function by simply fitting the expression in Eq. 5 to the data of bottom pressure head versus the average volumetric water content. The heterogeneous states of suction can be used to evaluate the hydraulic conductivity function of the soil as is described next.

PARAMETER ESTIMATION ANALYSIS

A parameter estimation algorithm based on Gauss-Newton method coupled with the line search technique as described by Dennis and Schnabel (1983) was developed and used to evaluate the parameters of the hydraulic conductivity function, a_2 and n_2. The test data needed in the analysis are the progress of bottom suction with time, or the time, t_e, needed to achieve specified values of bottom suction, s_i. The data obtained from the numerical simulation of the test for any estimates of a_2 and n_2, using program SEEP, are the progress of bottom suction with time, or predictions of the time, $t(c)$, needed to achieve specified values of bottom suction, s_i. These predictions are compared to the experimentally obtained times, t_e by calculating the normalized difference, HB_i, as

$$HB_i = (1 - \frac{t_c(s_i)}{t_e(s_i)}) \qquad (8)$$

The objective function, Q, needed to be minimized is defined as:

$$Q = 0.5 \sum_{i=1}^{i=m} (HB_i)^2 \qquad (9)$$

Table 1: Results of the Parameter Estimation Analysis.

Test No. (1)	a_2^i m^{-1} (2)	n_2^i (3)	a_2^f m^{-1} (4)	n_2^f (5)	N.O.I (6)	Q (7)
I	1.0	4.0	1.78	6.5	7	4.17×10^{-8}
	1.0	8.0	1.78	6.5	11	3.11×10^{-9}
	3.0	8.0	1.78	6.5	7	4.03×10^{-9}
	3.0	4.0	1.78	6.5	6	3.23×10^{-8}
II	1.0	4.0	2.0	7.0	7	4.60×10^{-8}
	1.0	8.0	2.0	7.0	8	5.37×10^{-8}
	3.0	8.0	2.0	7.0	7	4.6×10^{-8}
	3.0	4.0	2.0	7.0	6	4.52×10^{-8}
III	1.0	4.0	1.5	6.0	8	8.38×10^{-8}
	1.0	8.0	1.5	6.0	6	8.31×10^{-8}
	3.0	8.0	1.5	6.0	7	8.27×10^{-8}
	3.0	4.0	1.5	6.0	6	8.16×10^{-8}

where m is the number of specified bottom suction data. The parameter estimation algorithm minimizes the objective function and calculates improved values for a_2 and n_2 in each iteration. The analysis terminates when $Q \leq 1.0^{-7}$ or 15 iterations were completed.

The parameters a_2 and n_2 were estimated from the results of three numerical tests. The value of v used in Test I and Test II is $K_s/10$ and $K_s/4$ in Test III. The values of a_2 and n_2 used in creating the numerical test data are 1.78 and 6.5 in Test I, 2 and 7 in Test II, and 1.5 and 6 in Test III. Examples of the results of the parameter estimation analysis of test II in term of the predicted progress of the bottom suction with time and the hydraulic conductivity function are shown in Figures 8(a) and 8(b). The results of the parameter estimation analysis are summarized in Table 1. The initial estimates of parameters of a_2 and n_2 are listed in columns 2 and 3, respectively, and the final estimates of these parameters are listed in columns 4 and 5, respectively; the number of iterations, N.O.I, and the value of the objective function, when the stopping condition was met, are listed in columns 6 and 7, respectively.

CONCLUSIONS

It is obvious from the tabulated results and the agreement between the predicted and obtained progress of the bottom suction with time that the parameter estimation of the hydraulic conductivity parameters from the results of the controlled outflow test leads to a global minimum, which implies that the problem of

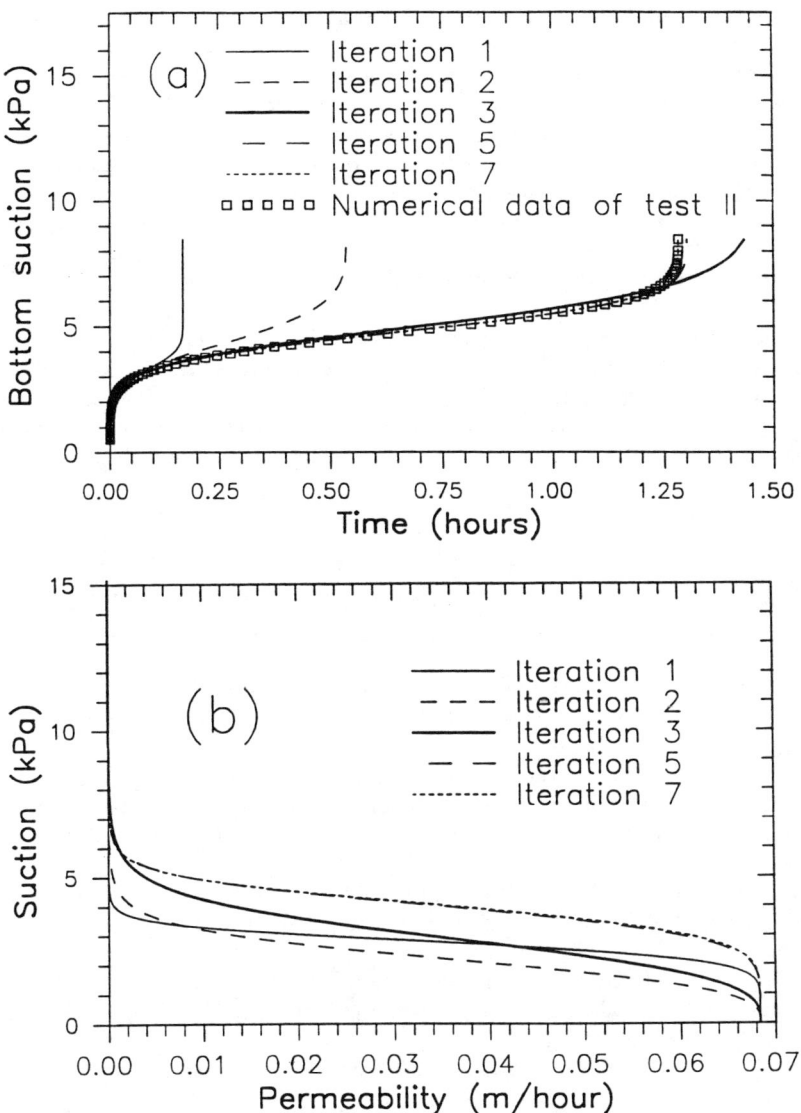

Figure 8: Results of the Parameter Estimation Analysis of Test II.

174 UNSATURATED SOILS

a non unique solution, often reported in the literature, was not encountered for these cases. The results at various iterations shown in Figure 8(a) indicate that the final portion of the test data plays a major role in the convergence to the correct solution. If only the first portion of the test data is used in the parameter estimation analysis, the problem of a non unique solution will be encountered since in that portion homogeneous states of suction exist within the sample that are independent of the permeability function, as shown clearly in Figure 4. This conclusion might give explanation for the non unique solution obtained by Kool et al. (1985) who concluded the need for the data in the final portions of the test results to reduce the probability of a non unique solution.

The presented analysis demonstrates the feasibility of the proposed flow pump technique for the determination of the constitutive properties for flow through unsaturated soils using the inverse problem solution approach. While the present analysis is restricted to the granular material with the negligible volume change and low suction, the concept can be expanded to the fine grained materials as well.

REFERENCES

(1) Abu-Hejleh, A.N. (1993). "Desiccation Theory for Soft Soils." Ph.D Dissertation, Civil Engineering Department, University of Colorado, Boulder, Colorado.

(2) Dennis, J.E. and Schnabel, R.B. (1983). "Numerical Methods for Unconstrained Optimization." Prentice-Hall Series in Computational Mathematics, Englewood Cliffs, New Jersey.

(3) Kool, J.B., J.C. Parker, and van Genuchten, M.Th. (1985). "Determining Soil Hydraulic Properties from One-step Outflow Experiments by Parameter Estimation:I. Theory and Numerical Studies." Soil Sci. Soc. Amer. J. 49:1348-1354.

(4) Olsen, R.E., and Daniel, D.E. (1981). "Measurement of the Hydraulic Conductivity of Fine-Grained Soils." Permeability and Groundwater Contaminant Transport. ASTM, STP, 746, 18-64.

(5) Manna, M. , D. Znidarcic, and Illangasekare, T.H. (1993). "Suction-Saturation Measurements in Soils Using the Flow Pump Technique." ASTM, Geotechnical Testing Journal (in print).

(6) Richards, L.E. (1931). "Capillary Conduction of Liquids through Porous Media." Physics, 1, 318-333.

(7) van Genuchten, M. Th. (1980). "A closed-form Equation for Predicting the Hydraulic Conductivity of Unsaturated Soils." Soil Science Society Amer. Proc., vol. 44, 892-897.

(8) Znidarcic, D. and Liu, J.C. (1989). "Consolidation Characteristics Determination for Dredged Materials." Proceedings of the 22nd Annual Dredging Seminar, Tacoma, Washington, pp. 45-65.

TRANSIT TIMES THROUGH COMPACTED CLAY LINERS

Priyantha W. Jayawickrama,[1] Associate Member, ASCE

Abstract

Transit time, or the time taken by waste liquid to move through the thickness of the liner, is a short-term performance parameter which can be used to determine the necessary liner thickness. In order to predict the transit time for a given liner configuration, one should be able to analyze the transient flow situation in which a wetting front of waste liquid moves through the initially unsaturated compacted clay mass. The analysis is further complicated by the fact that the liquid movement takes place along preferred paths and not uniformly throughout the entire cross-section. This paper reviews the current methodology for simulating the above flow condition and how they may be used for transit time prediction.

Introduction

Compacted clay is an essential component of the double-liner systems used in hazardous waste retaining facilities. Although, compacted clay soils have low hydraulic conductivities, they are, by their very nature, porous and permeable materials. Therefore, if a finite head of waste liquid is maintained over the clay layer indefinitely, the liquid will eventually seep through the clay. However, during the design and construction stages, the saturated hydraulic conductivity of the compacted clay can be controlled so that the amount of steady state seepage flux for a given liquid head is minimized. The above approach, which addresses the long-term performance of the liner, has been the basis for nearly all liner system designs thus far. A second performance criterion which may be considered in the design of liner systems is the *transit time*, or the time taken by the liquid waste to reach the liner base since the initial impoundment of the

---

[1] Assistant Professor, Department of Civil Engineering, Texas Tech University, Lubbock, TX79409-1023.

facility. The transit time for a given liner will depend on a number of parameters such as the hydraulic conductivity, the effective porosity, the liner thickness and more importantly the degree of defects caused by desiccation and/or poor quality control. A performance criterion based on the transit time may be used for the determination of the necessary bottom liner thickness as a function of the design life of the facility. Although the above approach was originally considered by the U.S. EPA, most liner design procedures used at the present time do not use transit time as a design consideration. However, a re-emergence of this concept is evident in the new regulatory requirements introduced by some states. This paper provides a critical review of the transit time prediction methodologies currently available.

Mechanism of Liquid Movement Through Compacted Clay

The data obtained from a number of research studies reveal that the hydraulic conductivity values determined based on field leakage rates are 10 to 1000 times larger than those measured in the laboratory (Daniel, 1984 and Boynton and Daniel, 1985). This large discrepancy has been attributed to the presence of various preferential flow paths or macro-pores that result from field construction practices. Other studies in which dyed permeants were used, provide data that confirm the existence of such macrofeatures in field liners (Brown *et al*, 1983, Brown *et al*, 1986 and U.S. EPA, 1989). These data suggest that the liquid moves downward through a system of macropores that is continuous within a single lift. On reaching the interface between lifts, the liquid spreads laterally before flow continues further into the next lift. Figure 1 illustrates the above flow mechanism.

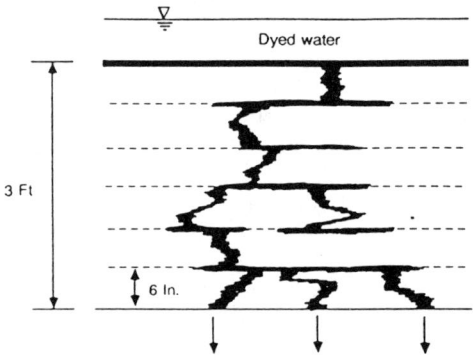

Figure 1.   Preferential Flow of Leachate

The presence of such a macro-pore structure significantly affects both the steady state leakage rate and the transit time through the liner.

Therefore, it is important that the methodology used for the prediction of transit times be capable of representing this flow mechanism that exists in field installed liners. Although, a significant number of different transit time prediction methods can be found in literature, the vast majority of them rely on equations that assume homogeneous soil conditions with uniform hydraulic properties (U.S. EPA, 1983; U.S. EPA, 1984). A comprehensive and detailed review of a number of these early prediction methodologies is available in U.S. EPA, 1988. This paper focuses on more recent methodologies which attempt to predict transit time through the liner based on flow through macro-features.

"Unifying Model" Model by Anderson et al., U.S. EPA, 1989

One of the earliest attempts to develop a preferential flow model to predict transit time through clay liners was presented by Anderson et al., (U.S. EPA, 1989). In this model each lift of compacted clay is represented as a slab with a population of evenly spaced, vertical, straight-sided cylindrical channels, all of which have the same radius. These large cylindrical channels represent the various defects in the clay. Additionally, the slab itself has a permeability, k_r which corresponds to the matrix permeability. Thus, the overall permeability of the compacted clay layer which consists of two separate components; the matrix permeability, k_r and macro-pore permeability, can be written as follows.

$$k = k_r + \frac{\pi R^4}{8 D^2} \quad \dots\dots\dots\dots\dots\dots\dots\dots\dots\dots\dots\dots\dots\dots\dots\dots\dots \quad (1)$$

It is demonstrated that, for pore size distributions typically found in clay liners, the matrix permeability is negligibly small when compared with the contribution from macro-pores. Therefore, in all subsequent analyses the matrix permeability term is not taken into consideration. The above model requires the user to provide the hydraulic conductivity and channel radius. The channel density, or the number of channels per unit area, necessary to yield the specified hydraulic conductivity is then calculated from the above equation.

A second part of the model deals with the *interlift flow*. According to this model, once the liquid reaches the interlift plane at the bottom of the cylindrical channel it will begin to spread radially outward. The radius of the wetted area, IR at any time t after the liquid had arrived at the bottom of the channel is given by the following equation.

$$IR = \sqrt{t \, K_h \, H} \quad \dots\dots\dots\dots\dots\dots\dots\dots\dots\dots\dots\dots\dots\dots\dots \quad (2)$$

The radius of the wetted area will gradually increase as a function of time(t), the interlift hydraulic conductivity (K_h) and the applied head(H). At any time a defect is found within this interlift flow radius, continuity is established between the two lifts and vertical downward flow towards the next interface will begin. The time taken to establish this continuity for a given pair of channels depends on the off-set distance between the two channels. This distance is assigned randomly by a computer program developed for this purpose. By using this program to generate connections between channels in adjacent lifts, the model can be used for the prediction of transit time through a liner consisting of a specified number of lifts. To accomplish this, the program must be run a large number of times and results recorded. Because the positions of the channels are determined randomly every time, each computation is an independent stochastic experiment which leads to a different set of effective flow paths and a corresponding transit time. The horizontal area of the liner used in each of the computations is 25 sq ft (5ft x 5ft). The above model is essentially a computer simulation of the actual leakage phenomenon. The model is not mathematically rigorous and involves a number of simplifying assumptions. Nevertheless, it incorporates a number of interesting features. The major shortcomings of the model with regard to its practical application may be as follows:

(1) Although an independent experimental study was conducted by the researchers, no attempt was made to collect data necessary for the validation of the analytical model.

(2) The model assumes that the vertical downward flow through the channels take place instantaneously. In other words, the entire delay in breakthrough of the liquids occurs as a result of the time taken for horizontal flow at the interface between different lifts. This assumption contradicts the observations later made by the researchers during the experimental study in which they concluded that *the rate of horizontal flow at interlifts far exceeds that of vertical flow within a lift.*

(3) No justification is given for the arbitrary selection of 25 sq ft (5ft x 5ft) as the liner area to be used in the computation of transit time. As the area increases the probability of establishing a flow path in a short time increases. Therefore the results obtained for a larger area can be very different. Furthermore, performing the computations for a larger area using the above simulation approach can be prohibitively expensive.

(4) For a given hydraulic conductivity, the number of channels calculated will depend on the assumed channel radius. In all calculations a channel radius of 250 μm is used. A different channel radius would have resulted in a different number of channels per unit area and

therefore a different channel spacing. Thus it is apparent that the assumed channel radius has direct bearing on the final results obtained for the transit time. Such sensitivity to an assumed input parameter is not desirable.

(5) Another input parameter that has important implications is the horizontal permeability, K_h of the interlift flow zone. Before the model can be used in actual design better guidance on the selection of a suitable value for K_h will be necessary.

"Macropore Model"; Benson (1989)

A second macro-pore flow model for transit time prediction was developed by Benson (1989). The term "macropore model" was used by the original researcher to distinguish this model from another model which assumed homogeneous hydraulic properties for the soil mass. The macropore model uses an approach which is very similar to the one previously described. Once again, the defects found within a lift are represented by a collection of vertical straight-sided cylinders. The interlift zone is modeled as a confined aquifer. The defects in the top lift (lift directly above the interface) are treated like a series of wells that drain liquid into the aquifer. Similarly, the defects in the lift below are wells that draw liquid out of the aquifer. Although the two models follow essentially the same approach, there are some fundamental differences between the two. In the *unifying model*, described previously, the macro-channels are initially empty. At time, t = 0 a known liquid head is established above the liner and downward liquid movement begins through the empty channels. The time at which liquid first arrives at the liner base was termed the transit time. In the *macropore model*, it is assumed that steady state flow conditions have already been reached throughout the entire system of macropores at time, t=0. The transit time, in this model, is the time taken by a particle of fluid to travel from the top of liner to the bottom under the already established steady seepage velocities. A second important difference is found in the idealization of the macropore system within a given lift. The *unifying model* assumes the radius of the pores and the spacing between them to be a constant. The *macropore model* treats these parameters as random variables which are distributed about a mean value. If the typical pattern for the distribution (such as normal, log normal, Poisson etc.) of the specific variable and its mean value is known, then appropriate values of the parameter can be generated by the computer.

As a first step, the parameters which define the macropore system within an individual lift must be determined. This is accomplished as follows: The total liner area is divided into a number of square partitions, each with an area equal to S^2. The number of channels per partition, N_m is assumed to be a Poisson Distribution. Then the number of channels corresponding to a given partition can be randomly generated if the mean value of N_m is known. For demonstration purposes the mean is assumed. Once a value for N_m has been selected, the mean pore radius can be determined by equating the total pore volume to the measured effective porosity as shown by the following equation.

$$N_m \ \pi \ r^2 \ L = n_e \ S^2 \ L \quad \dots\dots\dots\dots\dots\dots\dots\dots\dots\dots\dots\dots \quad (3)$$

Since the mean radius, r has been determined from Equation (3), the radius of each individual macro-pore can be generated by computer. The distribution of pore radii is assumed to be a normal distribution with a *small* standard deviation. Next, based on the observed hydraulic conductivity of the lift, a parameter known as the flow coefficient, J is calculated. J is assumed to be the same for all the pores.

$$\sum_{k=1}^{N_m} J \ \pi \ r_k^2 = K \ 1 \ S^2 \quad \dots\dots\dots\dots\dots\dots\dots\dots\dots\dots\dots\dots \quad (4)$$

As mentioned previously, the interlift zone is modeled as a confined aquifer with a series of recharge wells (macropores in the upper lift) and discharge wells (macropores in the lower lift). The interlift zone is characterized by its transmissivity, T. Thiem Equation is used to determine the hydraulic head at any point(i) in the jth interlift zone.

$$h_{i,j} = \frac{1}{2 \pi T} \sum_{k=1}^{TN_u + TN_l} Q_k \ \ln\left(\frac{r_{ki}}{r_k}\right) + W_j \quad \dots\dots\dots\dots\dots\dots \quad (5)$$

The flow rate, Q_k in k^{th} channel is given by the following equation.

$$Q_k = J \ \pi r_k^2 \ (\ H_{tk} - H_{bk} \ ) / L \quad \dots\dots\dots\dots\dots\dots\dots\dots \quad (6)$$

Equation (5) and (6) can now be combined to give the following.

$$h_{i,j} = \frac{\psi}{2 \, TL} \sum_{k=0}^{TN_u + TN_l} (H_{tk} - H_{bk}) \ J \ r_k^2 \ \ln\left(\frac{r_{ki}}{r_k}\right) + W_j \quad \dots\dots\dots \quad (7)$$

Equation (7) is written for each node within the interlift zone.

Similarly, equation of continuity can be written for each interlift layer in the following manner.

Equation (7) and Equation (8) result in a set of linear system of equations

$$\sum_{k=1}^{TN_u} J \ (H_{tk} - H_{bk}) \ r_k^{\ 2} = \sum_{k=1}^{TN_l} J \ (H_{tk} - H_{bk}) \ r_k^{\ 2} \ \dots\dots\dots\dots \quad (8)$$

which can be solved to obtain the hydraulic heads, H_{tk} and H_{bk} at the top and bottom of each channel. Once the above information has been obtained, the transit time computation can be started. Liquid may take many different pathways when it travels through the thickness of the liner. The pathway with the least time of travel provides the transit time. Each pathway will involve vertical movement from the top to the bottom of cylindrical channels as well as lateral movement within interlift layers. The time of travel in the vertical direction is given as by the following equation.

$$t_{ij} = \frac{L^2}{J \ (H_i - H_j)} \ \dots\dots\dots\dots\dots\dots\dots\dots\dots\dots\dots\dots\dots\dots \quad (9)$$

The time of travel in the lateral direction between channels k and l is calculated using the following *approximate* formula.

$$t_{kl} = \frac{r_{kl}^{\ 2}}{T/b \ (H_k - H_l)} \quad ; \ H_k > H_l \ \dots\dots\dots\dots\dots\dots\dots \quad (10)$$

The equations (9) and (10) above can be used to calculate the total travel time through any given pathway. Since there are an infinite number of such pathways, a computer simulation (Monte Carlo Simulation) is used in which the set of parameters which define a selected pathway is generated by *a random number generator*. The process is repeated a large number of times (1000 times) and the travel time computed for each random pathway. The minimum of these is reported as the transit time of the liner.

The following observations can be made regarding the *macropore model*.

1. The model does not simulate the transient flow situation that results from the initial impoundment of a liner which is originally at its compaction water content. Instead, it considers the time of travel through an already saturated liner in which steady state flow has been established.

2. The model has not been validated with test data.

3. This model is mathematically more rigorous than the *unifying model*, previously discussed. It better represents the variability of the many of parameters involved. However, computer time needed to complete sufficient number of simulations is very high because of the algorithms

used to randomly generate these variables. Therefore only a small section of the liner can be analyzed because, according to the researcher, *modeling an entire liner would be mathematically impossible.*

4. The model requires data such as, transmissivity of the interlift zone (T), the average number of macrochannels per partition (N_m), effective porosity (n_e) etc. For the purpose of demonstration, numbers were assumed for the above parameters. In order to use the model in practical applications guidelines for the selection of appropriate values for these parameters would need to be developed.

Model MACROFLOW; Jayawickrama (1990)

A third model for the computation of transit time through liners based on flow through macro-features was presented by Jayawickrama (1990). The model was developed from a study conducted at Texas A & M University. This research study consisted of two separate phases. The first phase involved a series of large-scale tests, in which two types of clay typically used in landfill liner construction was compacted into cylindrical tubs of 60 cm diameter and then tested for transit times and saturated hydraulic conductivity. Further details with regard to the experimental study may be found in the original reference (Jayawickrama, 1990). The second phase of the study consisted of theoretical developments.

The model MACROFLOW uses an approach which is distinctly different from the one used by the other two. The model considers transient flow through two separate systems; the macro-pore system and the micro-pore system. Unlike the two previous models, the MACROFLOW model treats the macro-pore system as a single continuum. Therefore, the model does not require information on macro-pore sizes, their variability or their spatial distribution. Instead, it requires global parameters that characterize the entire network of macro-pores as a whole. Such parameters are macro-pore hydraulic conductivity (K) and macro-porosity (θ_{ma}). The macro-pore network serves as the primary mechanism for flow through the liner. According to MACROFLOW model, at the time of initial impoundment a wetting front will begin to move through the macropore system. As it moves through the liner, lateral infiltration will take place from the macro-pore system into the surrounding clay matrix. Additionally, flow takes place through the intact soil matrix by direct infiltration from the top. However, liquid transfer through the intact clay occurs at a much slower rate. The governing equation for the macro-pore and micro-pore systems are as given below.

$$\int_0^H K\ dh = \left(\frac{dx}{dt}\right) \int_0^s \Delta\theta_{ma}\ ds + \Delta\theta_{mi} \int_0^x\int_s^x Q(t-t_r)\ drds \quad \dots\dots\dots\dots \quad (11)$$

$$\frac{\partial \theta_{mi}}{\partial t} = \frac{\partial}{\partial z} \left[K(\theta_{mi}) \frac{\partial h}{\partial z} \right] + S_u \quad \dots\dots\dots\dots\dots\dots\dots\dots\dots\dots\dots\dots \quad (12)$$

The transit time computation essentially involves the simultaneous solution of the two coupled governing equations. A numerical technique based on finite difference approach was used for this purpose. The model has the additional capabilities.

1. During the experimental study it was observed that the hydraulic conductivity of a single lift varied throughout its depth. It was concluded that the observed behavior resulted from the compaction energy being dissipated mostly near the top of the lift. The model has the capability accommodate this variation of hydraulic conductivity within the depth of a single lift.

2. Another observation made during the experimental study involved the time variation of hydraulic conductivity. It was evident that as the soil surrounding the macro-pores gradually reached saturation, the macro-pore conductivity decreased. The MACROFLOW model also has the capability to incorporate such changes in macro-pore conductivity. The changes in macropore-conductivity with degree of saturation in the intact matrix is given by the following equation.

$$\frac{K_0}{K} = 1 + \alpha \left[\frac{S_a - S_{ai}}{1 - S_a} \right]^{\gamma} \quad \dots\dots\dots\dots\dots\dots\dots\dots\dots\dots\dots\dots \quad (13)$$

The details concerning the development of the equation can be found in Jayawickrama and Lytton (1992).

The model described thus far pertains to flow within a single lift. In fact, the experimental program conducted as a part of this study was limited to flow through a single lift of compacted clay. The predictions from the model was compared with actual measurements of transit times and excellent agreement was obtained *(with maximum error less than 10%)*. The model was later extended to incorporate interlift flow. As explained earlier, direct connection between the macropore systems in the two adjacent lifts does not exist. Such connection is established by lateral flow that takes place along the interface. MACROFLOW model assumes that as soon as the wetting front through a single lift reaches the interface at the bottom of the lift, the liquid begins to spread laterally saturating a thin layer of material at the interface. This lateral flow will take place until a finite area within the interlift zone is saturated. Once this pre-determined finite area is saturated the necessary continuity is assumed to have been established, and the wetting front will resume its downward movement into the next lift. The

following observations are made regarding the use of the **MACROFLOW**
model in transit time prediction of liner systems.

1. The model does not require information on the distribution of
   individual macropore sizes, their variability or spatial distribution.
   Instead, it requires global parameters that characterize the entire
   macro-pore network as a single continuum. These parameters may be
   determined from large-scale infiltration test data by using a technique
   known as *system identification method.*

2. The validity of the model has been demonstrated *only* for single lifts of
   compacted clay. Before the model can be used as tool in an actual
   performance evaluation, better guidance will be needed in the selection
   of parameters that characterize interlift flow.

3. The model described above is a deterministic model which is applicable
   to a liner area which typical of the area covered by a large-scale
   infiltrometer test. The macro-pore hydraulic conductivity and the
   macro-porosity which will be used as input to the model are effectively
   spatial averages within this large area. The variability that is found on
   a larger scale, such as the differences in hydraulic conductivity
   measured from different infiltrometers can be incorporated into the
   model by using the *first order, second moment approach* in reliability.
   Such a model is being developed by the author at the present time.
   This approach eliminates the need for multiple computer runs unlike in
   the simulation approach used in the two previous models.

Conclusions

This paper provided a detailed review of three different transit time
prediction models, all of which were based on liquid transfer through
macro-pores. In modelling preferential flow, two general approaches have
been used. The first of these (used in "Unifying" and "Macropore" Models)
attempts to model liquid movement through discrete channels. Reasonable
assumptions are made regarding the size and spatial distribution of channels
and transit time calculated through a simulation process where relevant
parameters are generated through the use of a random number generator.
To simulate the infinite number of possible pathways calculations are
repeated a large number of times. The second approach (used in
MACROFLOW Model) uses a continuum approach where the entire macro-
pore network is treated as a continuum. The method makes no attempt to
represent the actual size and spatial distribution of macro-channels within
the compacted clay. Instead, it uses global parameters to characterize the
network of macro-pores as a whole. The model has the capability to

accommodate depth and time variation of the macro-pore conductivity. It also permits lateral flow from the macro-pores to the surrounding soil. The input parameters required by the model can be determined by double-ring infiltrometer tests. It should be also noted that all of the above models deal with the movement of the waste leachate itself and not the dissolved constituents in it which may interact with the soil.

Appendix I: References

Benson, C.H.,1989. A Stochastic Analysis of Water and Chemical Flow Through Compacted Soil Liners, *Ph.D. Dissertation*, Department of Civil Engineering, University of Texas, Austin, Texas.

Boynton, S.S., and D.E. Daniel, 1985. Hydraulic Conductivity Tests on Compacted Clay, *Journal of Geotechnical Engineering,* ASCE 111(4), 465-478.

Brown, K.W., J.W. Green and J.C. Thomas, 1983. The influence of Selected Organic Liquids on the Permeability of Clay Liners, *Proceedings, Ninth Annual Research Symposium on Land Disposal of Hazardous Wastes,* U.S. EPA, Cincinnati, OH, EPA-600/9-83-018:114-125.

Brown, K.W., J.C. Thomas and J.W. Green, 1986. Field Cell Verification of the Effects of Concentrated Organic Solvents on the Conductivity of Compacted Soils, *Hazardous Waste and Hazardous Materials* 3:1-19.

Daniel, D.E., 1985. Predicting Hydraulic Conductivities of Clay Liners, *Journal of Geotechnical Engineering,* ASCE 110(2); 285-300.

Jayawickrama, P.W., 1990. Liquid Transfer Through Preferential Paths in Compacted Clay, *Ph.D. Dissertation,* Department of Civil Engineering, Texas A & M University, College Station, Texas.

Jayawickrama, P.W. and R.L. Lytton, 1992, Conductivity Through Macropores in Compacted Clay, *Proceedings of the 7th International Conference on Expansive Soils;* 99-104.

U.S. Environmental Protection Agency, 1983, Lining of Waste Impoundment and Disposal Facilities, Document SW-870.

U.S. Environmental Protection Agency, 1984, Procedures for Modeling Flow Through Clay Liners to Determine Required Liner Thickness. EPA/530-84-001.

U.S. Environmental Protection Agency, 1988, Design, Construction, and Evaluation of Clay Liners for Waste Management facilities, U.S. EPA Risk Reduction Engineering Laboratory, Cincinnati, OH, EPA/530-SW-86-007F.

U.S. Environmental Protection Agency, 1989, Factors Controlling Minimum Soil Liner Thickness, U.S. EPA Risk Reduction Engineering Laboratory, Cincinnati, OH, EPA/PB91-191346-LDM.

Appendix II. Notation

D	=	channel spacing
H	=	applied head
H_{tk}, H_{bk}	=	hydraulic head at the top and bottom of channel
IR	=	interlift flow radius
J	=	channel flow coefficient
K	=	hydraulic conductivity of macro-pore system
K_0	=	initial macro-pore hydraulic conductivity
K_h	=	interlift hydraulic conductivity
$K(\theta_{mi})$	=	unsaturated hydraulic conductivity of matrix
L	=	lift thickness
N_m	=	number of channels per partition
$Q(t)$	=	lateral flow into surrounding soil
Q_k	=	flow rate in channel, k
R	=	channel radius
S	=	side length of square partition
S_a	=	degree of saturation
S_u	=	supply from macro-pore system
T	=	transmissivity of interlift layer
TN_u, TN_l	=	number of channels in upper and lower lifts
W_j	=	arbitrary constant
h	=	matrix suction of surrounding soil
$h_{i,j}$	=	hydraulic head at point, i in interlift layer, j
k, k_r	=	overall and matrix permeability
n_e	=	effective porosity
r	=	mean channel radius
r_k	=	radius of channel, k
r_{ki}	=	distance to point, i from channel, k
t	=	time
t_r	=	time of arrival of the wetting front at depth,r
x	=	depth of wetting front
α, γ	=	soil parameters
θ_{ma}, θ_{mi}	=	macro- and micro- porosity
ψ	=	+1 for recharge wells and -1 for discharge wells

HUD CONCERNS: POST-TENSIONED FOUNDATIONS IN TEXAS

Richard J. Sazinski, P.E., Ph.D., M.ASCE[1]

Abstract

An investigation was conducted of post-tensioned slab foundation systems in Texas pursuant to a request from the Headquarters Office's of the U.S. Department of Housing and Urban Development (HUD). Single-family, detached, residential type units, whose loans were insured by HUD, constructed over expansive soil, were of primary concern. The investigation, and subsequent report to HUD Headquarters, was conducted by HUD's Regional Structural Engineer (Denver). A trip to the Dallas/Ft. Worth, Texas area during March, 1990 was part of the investigation. The key points of the report are presented and include a discussion of the foundation system, in general, and the results of observations, discussions and inspections conducted during the area visit. Pre- and post- construction concerns are presented. The continued acceptance by HUD of post-tensioned slab foundation systems, designed and constructed per Post-Tensioning Institute (PTI) procedures (PTI, 1980), was recommended, but only under soils, design and inspection certification requirements, which are presented.

Introduction

The week of March 12-16, 1990 was spent by the writer in the Dallas/Ft. Worth area gathering data and information on single-family, post-tensioned slab foundation systems. This investigation was based on a request from HUD Headquarters to look into reported failures of this type of foundation system in certain areas of Texas. Based on these reported failures, HUD Headquarters was considering a temporary ban on their acceptance. It was the writer's goal to check out the problem and to compile a report for HUD Headquarters with recommendations.
----
[1]Regional Structural Engineer, U.S. Department of Housing and Urban Development (HUD), 1405 Curtis Street, Denver, Colorado 80202-2349.

The report summarized the foundation system, in general, and the results of observations, discussions and inspections conducted during the visit to the area.   The key points of the final report are presented.

Background

The scope of the report and investigation was limited to foundation systems defined and designed as Building Research Advisory Board (BRAB), Type III slabs (NAS, 1968), stiffened and reinforced, placed over expansive clay type soils in Texas.   Light frame residential construction for single-family residences, limited to two stories in height, was involved.   The typical type of post-tensioned slab foundation system under investigation is depicted in Figure 1.

The current design and construction method with which HUD requires compliance is the latest standard published by the Post-Tensioning Institute (PTI, 1980). HUD's Minimum Property Standards for Housing (HUD, 1984) references this standard.   Proper performance depends on site specific soils information (parameters for design), design by a licensed structural engineer, construction with adequate inspections, adequate site grading and drainage (during and after construction), and homeowner maintenance.

Soils & Design Parameters

Expansive clay type soils can cause two basic types of distortion modes depending on changes in the moisture content of the supporting soils.   The modes are depicted in Figure 2 and involve either a center lift ("center heave" or "doming") or an edge lift condition.   Center lift mode is usually a long-term condition which the slab slowly assumes over the years, whereas, the edge lift mode is more a seasonal or short-term condition.

Numerous factors can affect the overall successful design and performance of these types of slabs-on-grade (Figure 2).   Due to these factors, it is very critical to have the proper soils parameters defined along with proper engineering and construction.   A failure to properly account for any of the factors will ultimately lead to either a system failure or overall poor performance.

The PTI procedure (PTI, 1980) defines methods/ parameters to account for the climate (Thornwaite Moisture Index - I_m) for predicting the edge moisture variation distance (e_m).   The procedure is an approximation with a stipulation statement that "the value of e_m to be used in structural design calculations should be provided

in the soils investigation report submitted by the geotechnical engineer." The soils engineer should define e_m.

The PTI procedure (PTI, 1980) defines a method that
may be used by a geotechnical engineer to estimate the
differential soil movement (y_m) for both edge and center
lift modes. The e_m value must be known along with other
soils information usually obtained by laboratory testing
of soils samples obtained from the site (type and amount
of clay material, initial wetness, depth of active zone
and velocity of moisture infiltration or evaporation).

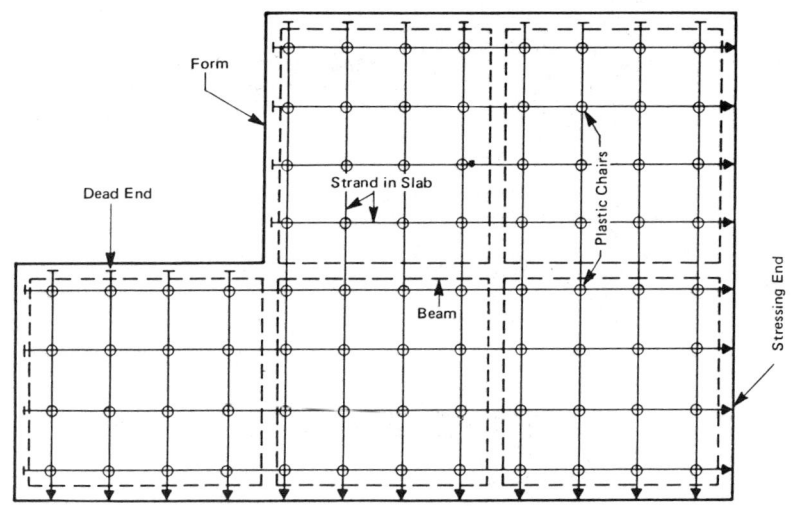

Typical Plan

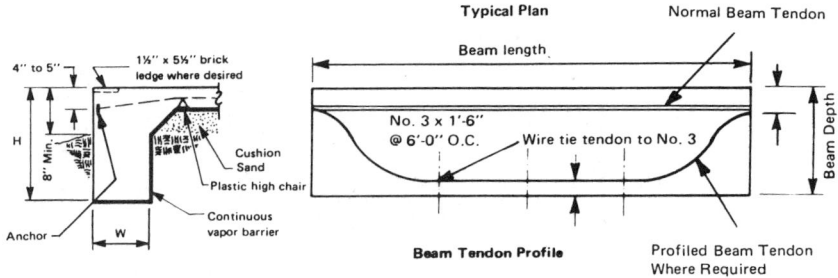

Tendon elevation and typical section

Figure 1. Typical Post-Tensioned Slab Foundation System
(PTI, 1980)

FACTORS AFFECTING SLAB DESIGN AND PERFORMANCE

1) <u>CLIMATE</u>: Semi-arid areas are more critical (i.e., periods of rainfall followed by extended dry periods).

2) <u>SWELLING MODE</u>: Edge-lift vs. center-lift.

3) <u>EDGE MOISTURE VARIATION DISTANCE</u>: (e_m)

4) <u>DIFFERENTIAL SOIL MOVEMENT</u>: (y_m)

5) <u>OTHER FACTORS</u>: (not related to climate)

   a) pre-vegetation,
   b) fence lines, trails and tracks,
   c) slopes,
   d) cut and fill areas,
   e) grading and drainage,
   f) time of year of construction (dry or wet period),
   g) post-construction practices (i.e., changes/additions made by homeowner relative to landscaping, watering, grading, drainage, etc.).

6) <u>STRUCTURAL PARAMETERS</u>:

   a) slab length (fixed by building plan),
   b) stiffening beam spacing, usually 10 to 20 ft (3 to 6 m),
   c) loading,
   d) depth of stiffening beams, and
   e) stiffening beam width, usually 8 to 12 in (20 to 30 cm).

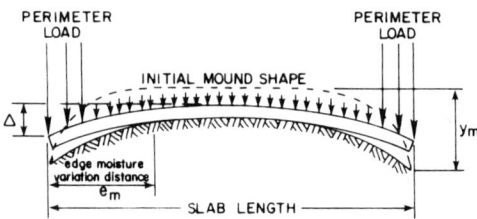

CENTER LIFT

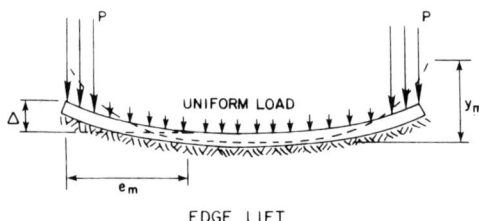

EDGE LIFT

Soil-structure interaction models

Figure 2.   Behavior and Factors Affecting Slab Design and Performance (PTI, 1980)

With e_m and other soils parameters, PTI tables are available for predicting y_m for edge and center lift modes.
This is dependent on correcting site conditions so that
soil moisture conditions are affected by climate only.

The uncertainty as to the adequate handling of variable site conditions can have a significant affect on the
soil moisture conditions and on the performance of the
stiffened slab. If these conditions are not properly
addressed, then soil movements larger than those resulting from climatic influences (changes) alone can occur.
The key point to realize is that some degree of engineering judgment must be exercised by the geotechnical
engineer in properly specifying y_m, for both edge and
center lift modes, so that any potential influence of
other factors, other than climate, is also accounted for.

Other factors, not related to climate, may induce
soil movements larger than those resulting from climatic
influences alone. Such factors may include pre-construc-
tion vegetation, fence lines, trails, tracks, slope of
the site, cut and fill areas, grading and drainage, time
of year of construction (dry or wet period), and post
construction activities (i.e., changes/additions made by
homeowner relative to landscaping, watering, grading,
drainage, etc.). The importance and critical nature of
these factors cannot and should not be overlooked.

Structural Parameters & Design

PTI procedures (PTI, 1980) require all key soils
parameters (soil bearing pressure, e_m, y_m, etc.) to be
properly and adequately defined. Following that, specific structural design formulas and procedures are
defined for the design of these types of stiffened slabs
on expansive soil. A key procedure, however, in this
whole design scheme is that two types of swelling modes
must be designed for (center and edge lift). Both design
modes must be checked and adjustments made as necessary.

Once the design is completed, the final requirements
must be defined on a foundation plan (with sections and
details) along with a list of general notes. The items
noted in Table 1, as a minimum, should be included on the
foundation plan and are excerpted from Sections 8 and 9
of the PTI design and construction manual (PTI, 1980).

Construction Inspections

Supervision over the construction of this type of
stiffened slab is usually more critical than for a con-
ventionally reinforced stiffened slab. The engineer who
designed the system should submit certification(s) stat-

UNSATURATED SOILS

Table 1. Foundation Plan Requirements

---

1) Lot(s), block(s) and subdivision designation. Signature and stamp (seal) of a professional (structural) engineer (P.E.) registered/licensed in state where unit is constructed.

2) Required concrete strength at 28 days and required concrete strength at time of stressing (usually 2000 psi/13.8 MPa) and time frame for allowing stressing (usually 3 to 10 days after pouring concrete).

3) Initial jacking force, slab tendon spacing, and other required sections and details. Special notes for post-tensioning materials and installation [per Section 8.3 of PTI manual (PTI, 1980)]. Tendon and reinforcing steel/mesh placement tolerances [per Section 8.4 of PTI manual (PTI, 1980)].

4) Site grading and drainage requirements [fill certification if required, minimum grades away from foundation, set exterior grade no more than 8 in. (20 cm) below top of slab, etc.].

5) Note referring to Soils Report which formed the basis for the foundation design [list soils foundation design parameters; e_m and y_m for both edge and center lift, allowable soil bearing pressure, slab subgrade friction coefficient].

6) Inspection and certification requirements for placing, concreting and tensioning per Sections 8 and 9 of PTI manual (PTI, 1980).

---

ing that the placing, concreting and tensioning were done in accordance with the design. The only possible way for the engineer to provide a valid certification(s) is to conduct adequate inspections of on-going construction.

A four part inspection procedure should be followed according to Sections 8 and 9 of the PTI design and construction manual (PTI, 1980) and is as follows:
(1) Inspection prior to concrete placement to check forming, geometry, and tendon and reinforcing steel type and placement.
(2) Inspection during concrete placement to verify concrete mix and proper consolidation; no damaging displacements of tendons, reinforcing steel or end anchorages; no entrance of cement paste into anchorage and proper curing techniques utilized.
(3) Inspection during stressing to verify proper forces, gauge pressures or elongations. All hydraulic jacks utilized are properly calibrated. Construction loads on concrete minimized until slab stressed. Stressing conducted between 3 and 10 days after pouring with a minimum concrete strength of 2000 psi (13.8 MPa).
(4) Final inspection to ascertain that all strand ends properly cut or burned off, all stress pockets grouted within 7 days of final stressing operations, and that any problems, repairs, etc. which were noted previously, have been completed or cleared.

Post-Construction Homeowner Involvement

When construction and lot grading are completed, a key period is entered, homeowner involvement. A house with a properly designed and constructed foundation and interior, along with proper initial exterior grading and drainage, can still be damaged if good landscaping and maintenance practices are not followed by the homeowner.

Landscaping includes the shaping of the land and moisture control as well as planting of vegetation. Good landscaping and maintenance are essential to reducing the amount of water that infiltrates the ground causing expansive (swelling) soils to swell. If landscaping is carefully planned and maintained, damage from swelling soils can be minimized.

When expansive soils become wet and cause damage due to lack of timely maintenance by the homeowner, problems can be caused by:
(1) neglecting slope maintenance for good drainage,
(2) neglecting to clean gutters and downspouts,
(3) over-saturating foundation soils by excessive land-scape watering,
(4) planting trees, shrubs, and flowers too close to the foundation walls,
(5) changing lot drainage pattern by constructing patios, fences or obstructions that dam and pond water, or
(6) failing to seal old construction joints and cracks that develop over time in exterior flat work (avenues for water intrusion).

Recent Observations (March, 1990)

The writer spent the week of March 12-16, 1990 in the Dallas/Ft. Worth area. Data and information was gathered on single-family, post-tensioned slab foundation systems. Field inspections were conducted as well as discussions with numerous parties (builders, university staff, engineers, repair contractors, etc.). Table 2 lists parties contacted either prior-to, during, or after the trip to the Dallas/Ft. Worth area. Discussions were either face-to-face, via telephone contact, or both, and provided excellent background information and opinions.

Design Comments & Observations

(1) PTI design procedures (PTI, 1980) are being utilized, for the most part, by the consulting engineering community, however, several designs are being utilized which are not fully per the requirements or recommendations of the PTI procedures (PTI, 1980).
(2) Some design engineers use the PTI procedures (PTI,

Table 2. Personal Contacts (*Telephone Contact Only)

NAME	TITLE/AFFILIATION	LOCATION
1) Les Lane	Const. Srvcs. Mgr., HOW	Irving, TX
2) I. Shahparasti	Technical Rep., HOW	Irving, TX
3) Ken Garrett	Const. Investigator, HOW	Irving, TX
4) Tom Hollenbach*	HOW	Wash., D.C.
5) Robert Lytton*	Prof., Texas A&M Univ.	C. Station, TX
6) Henry Hadnot*	HUD-Architect	Houston, TX
7) Ken Crandall*	HUD, Chief Architect	Wash., D.C.
8) Thomas M. Petry	Assoc. Prof., Univ. of Texas	Arlington, TX
9) W. Kent Wray*	Prof., Texas Tech Univ.	Lubbock, TX
10) John H. Matthys	Prof., Univ. of Texas	Ft. Worth, TX
11) Victor Walters	HUD-Architect/Engineer	Ft. Worth, TX
12) Woody Hickman	President, Hallmark Homes	Irving, TX
13) Walt Patterson*	Energy Ext. Srvcs., U. of TX	Arlington, TX
14) C. Freyermuth*	Manager, PTI	Phoenix, AZ
15) Stone Stephens	Claims Rep., HOW	Irving, TX
16) Robert W. Brown	Pres., Brown Fndtn. Repair	Garland, TX
17) Robert L. Brown	Brown Fndtn. Repair, Inc.	Garland, TX
18) Doug Stines*	Structural Engineer	Denver, CO
19) Ron Reed*	Reed Engng./Soils Mechanics	Dallas, TX
20) Bob Henry	Bowen Structures	Bedford, TX
21) Barbara Gibson	HOW-Underwriting	Irving, TX
22) Ed Snider	Technical Rep., HOW	Irving, TX
23) Mark Lampton	Claims Rep., HOW	Irving, TX
24) Del Wachtal	HUD-Complaint Officer	Ft. Worth, TX

1980) with a more conservative edge lift allowable deflection criteria (1/3400) than the permissible edge lift deflection ratio defined by PTI (1/1700).

(3) Edge lift may always control the design due to the more stringent allowable deflection criteria compared to center lift (1/1700 vs 1/360).

(4) Some engineers may only do a center lift design and not perform the edge lift design check, which may usually control the design of this type of system.

(5) Obtaining a foundation design for this type of system at a "shopping mall" has been reported (i.e., obtain a somewhat non-site-specific foundation design from any store providing generic foundation plans).

(6) Very little or no soils information is apparently generated for each lot in a subdivision. Soils tests and/or information on each lot may be necessary to define site-specific foundation design parameters for this system.

(7) Several Soils Reports reviewed appear to show that the geotechnical consultants define the design parameters for this system (e_m, y_m, etc.) on a subdivision scale (i.e. same parameters for each lot).

(8) Several geotechnical consultants have experienced or seen greater differential soil movements (y_m) than those predicted via the PTI procedure which involves the use of various PTI curves and tables (PTI, 1980). NOTE: The PTI procedure (PTI, 1980) does comment that extremely active clays have been reported which

may generate larger values of edge moisture variation distance (e_m) and consequently larger values of vertical movement (y_m) than reflected by the curves and tables. For this reason, the PTI curves and tables should be used only in conjunction with a site-specific soils investigation from a geotechnical engineer knowledgeable of local soil conditions.

Construction Comments & Observations

(1) Poor quality control on foundation construction is in evidence in some cases (low quality of inspections or no inspections at all).

(2) In some cases, concrete quality on the job is downgraded by adding water to the ready-mix truck.

(3) Some tendons may not be pulled or stressed. Tendon wrapping may be damaged or torn on the job site.

(4) Slab foundations cast at the end of a lengthy dry period (i.e., summers-normal construction season) experience greater uplift around the edges (i.e., edge lift).

(5) Tendons unchaired prior to casting concrete. Pull-throughs of tendons occur due to missing anchor plates prior to casting concrete.

(6) Very little or no prewetting of soil is done prior to casting concrete during a dry construction period. Adequate prewetting may be difficult to achieve due to dense soils and very slow moisture migration through the soils.

(7) Construction on cast concrete portions of slab foundations, prior-to-stressing, usually involves significant framing and in some cases, brick veneer. Minor framing is acceptable, however, loads should be minimized before stressing.

(8) Stressing/pulling of tendons is usually completed 7 to 9 days after the concrete is cast. Subcontractors do stressing/pulling with poor control, in some cases, and with no or somewhat outdated jack calibration sheets. No engineer is on-site to perform inspections of the stressing/pulling operations.

(9) "U" shaped layout and pulling of tendons has been reported (not recommended by the PTI).

(10) Very few problems have been reported with the quality of the prestressing material (tendons, cables).

(11) Some pretreatment of soils has been utilized by some builders to stabilize the expansive soils via chemical injection prior to foundation construction. Good results have been achieved in several cases.

(12) In some cases, no engineer's certifications are generated as to whether proper foundation installation is ever achieved.

(13) Poor grading and drainage, on several lots inspect-

ed, was evident throughout the construction phase.
(14) Several problem lots inspected, involved placed
engineered fill (inadequate compaction), cut and
fill, and/or poor foundation subgrade preparation.

Post-Construction Comments & Observations

(1) Homeowners create problems via changes to grading,
drainage, added landscaping and poor maintenance.
(2) Post-tensioned systems appear to function properly
if correctly designed and installed.
(3) Problems, if any, appear to show up around one to
three years after the unit is constructed.
(4) Very little cracking is evident in the slabs, however, since this system is more flexible, more
movement can occur with less cracking.
(5) Floor plans resulting in a foundation layout that is
not typically rectangular (i.e., L, T, U, H or I
shapes) seem to result in more problems.
(6) Problems around foundation edges seem more prevalent
(area most susceptible to soil moisture change).

Conclusions & Recommendations

Based on the writer's observations and discussions,
the continued acceptance of post-tensioned slab foundation systems constructed over expansive soils in Texas
was recommended. However, a more strict set of conditions was recommended for their continued acceptance by
HUD. A certification process was recommended as a possible means to assure proper design and construction.

Three certification sheets would make up the certification process and would address geotechnical, design
and construction inspection areas which are all critical
for proper system performance. Figure 3 depicts the
geotechnical engineer's certification sheet which could
be submitted and attached to a soils report for the lot
in question.

Two additional certification sheets, similar to
Figure 3, were also recommended but are not shown; one to
assure proper design and the other to assure proper
construction inspection. These two additional certifications would be completed by a registered/licensed professional (structural) engineer (P.E.). The design certification sheet, attached to the unit foundation plan and
referencing the soils report, would verify design
considerations per PTI procedures (PTI, 1980). The construction inspection certification sheet would verify
construction in compliance with the unit foundation plan
and would address construction inspection items outlined
in the PTI procedures (PTI, 1980).

SOILS/GEOTECHNICAL ENGINEERS CERTIFICATION

(Post-Tensioned Slab Foundation System)

SUBDIVISION NAME:_____

ADDRESS:_____ LOT & BLOCK:_____

CITY:_____ STATE:_____ ZIP:_____

CERTIFICATION:

I,_____, Registered/Licensed Professional
(geotechnical/soils) Engineer (P.E.), to the best of my knowledge, belief
and professional judgment, do hereby certify that:

(1) the soils report/investigation entitled _____
    and dated_____for the lot/block/subdivision referenced
    above, addresses the applicable topics which are covered in Section
    3.4 of the Post-Tensioning Institute (PTI) publication entitled
    "Design and Construction of Post-Tensioned Slabs-On-Ground" dated
    1980.

(2) the foundation design parameters recommended are as follows:

    *edge lift: (e_m)_____(ft)
                (y_m)_____(in)

    *center lift: (e_m)_____(ft)
                  (y_m)_____(in)

    allowable soil bearing pressure_____(psf)

    slab subgrade friction coefficient_____.

*NOTE: The parameters recommended for design take into account not only
       the soil & climate but also possible affects of pre-vegetation,
       fence lines, trails, tracks, slopes, cut & fill areas, grading &
       drainage, possible time of year of construction & possible post-
       construction changes.

I am aware of applicable sanctions provided in Title 18 U.S.C. 1010 and
hereby declare that the foregoing statements are not false, fictitious,
or fraudulent. I am also aware that HUD will not accept future statements
from me if any of the foregoing statements are false or incomplete.

Signed:_____
Professional's Name:_____
Business Address:_____
_____
Telephone Number:_____   (SEAL)

Figure 3. Geotechnical Engineer's Certification Sheet

All certification sheets were recommended to be
fully completed or risk rejection. It was also recommended that this certification process be monitored for a
five year period to check its effectiveness.

Relative to proper grading and drainage for construction, current HUD requirements (i.e., exhibits,
slopes, grades, etc.) are addressed under HUD handbook
requirements (HUD, 1990). Appendix 8 of this handbook

(HUD, 1990) defines grading and drainage guidelines.
Any new units utilizing this system would typically need
to comply with HUD's grading and drainage guidelines.

Relative to proper homeowner awareness and/or maintenance (i.e., grading, drainage, landscaping), it was
recommended that HUD consider a "Booklet For Homeowner
Maintenance", or other similarly entitled document, to be
developed by HUD's Policy Development and Research (PD&R)
Section in Washington, D.C. It would define proper maintenance procedures for homeowners relative to grading,
drainage and landscaping in areas with expansive soils.
The booklet would be developed for nationwide use and
awareness. Some similar booklets have already been
developed but only on a state-by-state basis. Such a
booklet would be provided to the homeowner at closing for
home purchases utilizing HUD mortgage insurance in an
area of known expansive soils.

The views presented in this paper are those of the
writer as a HUD employee and an individual and do not
necessarily represent the views of the United States
Government, in general, or those of the Department of
Housing and Urban Development (HUD), in particular.

Acknowledgments

The writer wishes to express gratitude to a
Mr. Kenneth L. Crandall (Chief Architect, HUD
Headquarters, Office of Insured Single Family
Housing-HSIDV, Washington, D.C.) for comments,
suggestions, and encouragement in writing this paper.

References

National Academy of Sciences (NAS), 1968, Building
    Research Advisory Board (BRAB), "Criteria for
    Selection and Design of Residential Slabs-on-
    Ground," Report No. 33 to the Federal Housing
    Administration, Publication 1571, Washington, D.C.

Post-Tensioning Institute (PTI), 1980, "Design and
    Construction of Post-Tensioned Slabs-On-Ground,"
    PTI, Phoenix, Arizona.

U.S. Department of Housing & Urban Development (HUD),
    1984, "Minimum Property Standards for Housing," HUD
    Handbook 4910.1, Washington, D.C.

U.S. Department of Housing & Urban Development (HUD);
    March, 1990, "Architectural Processing and
    Inspections For Home Mortgage Insurance,"
    HUD Handbook 4145.1 REV-1, Washington, D.C.

PERFORMANCE OF HORIZONTAL AND VERTICAL BARRIERS

Malcolm Steinberg, P.E., F.ASCE[1]

Abstract

Horizontal and vertical barriers have been used on transportation facilities to control expansive soils and minimize their damages. The effort is to minimize moisture change in the soil and minimize its often destructive volumetric change. The use of geomembranes, a waterproofed fabric, on over one hundred projects provides this paper's focus. There are significant indications that geomembranes can frequently minimize the damages caused by the expansive soils.

Introduction

Expansive or swelling soils cause annual damages in the United States exceeding $10 billion and to the Chinese railroad system 100,000,000 Yuan (Chinese dollars). Early efforts to minimize damage to pavement structures due to expansive soils were attempted in Texas, Mississippi and South Africa (ponding), Europe and Arizona (electroosmosis). Texas, Oklahoma, Arkansas, New Mexico, Utah, and Idaho attempted lime stabilized subgrade damage mitigation. Utah, New Mexico, and Texas experimented with subexcavation of expansive soil. Colorado, Utah, Idaho, Mississippi, and South Africa utilized asphalt treatment of expansive soil subgrades. It was the asphalt treatment efforts of Colorado that led to the first use of geomembranes, a waterproofed engineering fabric, to create a moisture migration barrier for expansive soils. (Steinberg, 1980 & 1992)

---

[1]   Adjunct Research Engineer, University of Texas at El Paso, El Paso, TX

Horizontal Barriers

The Colorado Department of Transportation in the early 1980's used a DuPont donated Typar on an 800' (290m) highway section between 6" (150mm) layers of sand over a Mamcos shale. A 1967 field inspection found the sand to be wet, concluded the method unsatisfactory, and since that time has been blasting and recompacting the shale. (Personal Communication Barrett 11/92)

The Texas Department of Transportation (TexDOT) used the 600' (182 m) of Typar on General McMullan Drive in southwest San Antonio, an urban arterial whose pavement had experienced severe distortion. The fabric exhibited no tares after base placement. An asphalt surface treatment and asphaltic concrete pavement followed. No repair work was needed until 1989 when pot hole filling, rotomilling and overlaying took place. (Personal Communication G.K. Hewitt 12/93)

Two San Antonio streets rehabilitated by TexDOT used a horizontal geomembrane placement. Colorado Street on the city's west side has expansive clays and in 1988 used 2000' (600 m) of a Mirafi geomembrane. Five years later the section remains in good condition, while an adjacent section without fabric has a distorted pavement surface. Another recent project, Guadalupe Street, with a clay subgrade used 1000 to 2000' (300 to 600 m) of geomembrane and the pavement remains smooth. (Personal Communication R.E. Magers, 1/93)

IH 80 Wyoming

The Wyoming Department of Transportation (WYDOT) first used the fabric on IH 80 placed horizontally across the existing pavement over a Mancos shale and covered with a non-expansive soil 6" (150 mm) deep. Base placement and an asphaltic concrete pavement followed. The 11.3 miles long (16.9 km) project, used 108,854' (33494 m) of fabric and was completed in 1983. In 1993 the road looks good and the fabric's considered to have been helped. The project was the first of over 50 using geomembranes by the Wyoming DOT. (Personal Communication Hager, 1/93)

U.S. 395 Nevada

In 1990 Nevada DOT used a geomembrane horizontally on a US 395 rehabilitation project. The fabric was placed horizontally following 5' (1.6m) of overexcavation replaced by non expansive material. The 3000' to 4000' (914 to 1220 m) fabric section is working fine. The department feels removing and replacing the expansive material alone did not work well. (Personal Communication Cochran, 1/93)

U 10 Utah

A five mile (7.5 km) U10 rehabilitation project used 114.222 S.Y. (95500 m) of Typar 3401 bid at 59¢/SY. (0.836/m²) tacked to the existing pavement. UDOT noted no swelling or heaving has occurred, a cost saving was achieved and public criticism was eliminated. (Spensko)

IH 40, Arizona

The IH 40 Arizona Department of Transportation (ADOT) 10 mile (15km) rehabilitation project completed in 1992 between Needle Mountain Road and SH 95 used 4400' (1340m) of geomembrane placed horizontally. SR 347 used a mile (1.6 km) of geomembrane when heavy rains created construction subgrade difficulties. Previous pavement distortions have not reappeared. ADOT is planning similar placement on SR 73. (Personal Communication Beekman, 2/93)

IH 20 Mississippi

Mississippi Department of Transportation reported the use of geomembranes on IH 20 in Scott County. The fabric was placed on 13 locations of the east bound lane and 14 locations on the westbound lane, where the pavement showed the most heaving. A Phillips Petromat MBII was used horizontally, 9' (3m) from the centerline of the lane in each direction extending to 6' (2m) beyond the ditch lines. The geomembranes totalled 142,820 SY (119,398 m²), bid at $2.33/SY (.836 m²), costing $324,198.80. Construction was completed in 1990 and roughness testing was inconclusive. (Browning, 11/92)

China's Railroad Lines

One fifth to one sixth of the Chinese rail system crosses expansive soils, Geomembranes, a polypropylene fabric, 0.95 mm thick, coated with emulsified bitumen, were used horizontally, on the Tai-Jao line in northern Shanxi Province. The fabric was placed between two sand layers 5 to 7 cm thick and the work was completed in November 1989. Displacement measurements were taken every two months, averaging vertically 5 mm that gradually decreased, and none laterally. Boiling and mudpumping no longer occurs on the geomembrane sections while severe damages continue on the untreated ones. They conclude

that compared to usual repair and maintenance costs the
geomembranes remedial cost is very little. (Wu 1992)

Bureau of Indian Affairs

Six Bureau of Indian Affairs reservations used
geomembranes horizontally. Two on Navajo lands used
Mirafi MCF 1212, and two used Typar 3353 with
satisfactory results being reported on all (personal
communication - Cotecson). Other BIA offices reported
two other fabric projects one of which WAS served well
for eight years.

IH 40 New Mexico

A recent IH 40 rehabilitation contract west of
Albuquerque used a Phillips MB fabric horizontally after
2' (0.6m) of subexcavation and then turned 2' vertically.
The vertical trench was backfilled with native material.
(Personal Communication - Lueck 1992)

VERTICAL BARRIERS

US 12 South Dakota

Probably the first placement of vertical moisture
barriers using geomembranes took place on South Dakota's
U.S. 12 in 1965. The fabric was placed 4'(1.3m) deep
along both roadway shoulders. The assessment indicated
that the fabric helped reduce the heave but if placed
twice as deep it would have been more effective. Current
SD DOT practice is to undercut up to 6' (2.m) on
Interstate Highways. (McDonald, 10/73) (Personal
Communication - D. Anderson, 11/92)

IH 410 Texas

The rehabilitation of IH 410, a loop around southwest
San Antonio, used Texas' first deep vertical fabric
moisture barrier (DVFMB) in 1978. The original
construction was a hot mix asphalt concrete pavement over
a flexible base and an expansive subgrade. In the Valley
High Drive underpass area the mainlands in the cut
section had severely distorted pavements. The
rehabilitation contract included 4978' (1531 m) of
geomembrane bid at $20/Ft. (0.348 m), placed 8' (2.5 m)
deep, and lapped 2' (0.6 m) to the paved shoulders,
typical for all later Texas DVFMB projects. The 8' (2.5
m) depth derived from previous studies that indicated the
zone of activity for moisture change ranged from 6' to 8'
(1.8 to 2.5 m). Some sliding of the trench walls took

place during the backhoe operations. It was solved by the use of a sliding shoring pulled by the backhoe. The testing with the profilometer, photologging and moisture sensors all indicated that the fabric section was providing a better roadway than the adjacent southbound lane without geomembrane. (Steinberg, 1985)

IH 37 Texas

IH 37 in southeast San Antonio, constructed in the 1960's, had reinforced concrete pavement placed over a cement stabilized base, a lime stabilized subgrade, and beneath it a swelling clay. This is the same section as the projects on US 281 and IH 10. Maintenance in the IH 37 2 mile (3 km) cut section varied from $50,000 to $100,000 annually. A 1981 rehabilitation contract removed the median ditch, sloped the pavement to the outside shoulders where the DVFMB was installed. An asphalt levelup and surface followed. The Typar T063 fabric was bid at $21/SY (0.836 m^2) for the 21483 SY (17960 m^2). A trenching machine was used for the excavation with a daily goal of 600' (184 m). Greater moisture changes took place outside the fabric protected subgrade than inside, while profilometer readings indicated a smoother riding surface on the geomembrane roadway compared to the adjacent control section. In 1984 the north control received a level up and its reading of a smooth ride exceeded the fabric segment. Through 1993 no pavement levelups have been required and is reported to be doing 'pretty well'. (Personal Communication R.E. Magers, 2/93)

Another San Antonio freeway, US 281 on the city's northside, received rehabilitation in the early 1980's. Constructed in the 1970's the southbound lane in a cut section began to have considerable pavement distortion. The contractor bid the 4705 SY (3933 m^2) DVFMB, a Mirafi MCF 500 at $3.18/SY (0.836 m^2), and used a backhoe for the excavation. Profilometer readings indicated the fabric protected lane had a smoother ride than the unprotected adjacent northbound lane. The 1993 advisory reported the northbound lane received a resurfacing and the fabric section is doing fine.

IH 10 on San Antonio's east side between Pine and Amanda Streets was built in the late 1960's. The depressed section had pavement distortion beginning shortly after construction. A 1985 contract provided for 24745 SY (20687 m^2) of DVFMB along both shoulders of both mainlands. The contractor bid a Mirafi MCF 500 & 140 N at $15/SY (0.836 m^2) and used a trenching machine for

excavation. When some sliding took place the trench was
moved further away from the shoulders. The tie to the
paved shoulder was made by placing fabric horizontally,
and covered with subgrade for protection from ultraviolet
ray deterioration. In 1993 the fabric protected lanes
provided a smoother ride than the control sections, and
the project was holding up well.

US 87 is in southeast San Antonio constructed as a
four lane divided highway over an expansive black clay
subgrade. The initial design was a lime stabilized
subgrade, a course of flexible base, an asphalt seal and
asphaltic concrete pavement. The 1987 widening and
reconstruction of a heaving pavement included 45000'
(13846 m) of DVFMB bid at $13.50/LF (0.304 m). The Typar
EVA fabric was placed at an average daily rate of 800'
(246 m) without encountering any problems during, after
or to this date. (Personal Communication R.E. Magers
2/93)

IH 10 Texas

Two projects in west Texas on IH 10 had DVFMB
placement to control expansive soils. The work in
Hudspeth County was constructed in the 1960's, and
included a 5' (1.5 m) subgrade undercut of the expansive
bentonitic clay, replaced with a low PI inorganic
material. The finished pavement system included a
flexible base and an asphaltic concrete surfacing. The
roadway remained stable for several years then distorted.
Remedial work in 1984 included 50098' (15270 m) of DVFMB,
bid at $13.28/LF (0.3 m). The contractor used a Vermeer
trencher in the Typar 3358 placement that averaged
500'/day 160 M) marred by caving and shoulder loss, where
underdrains were placed on the original contract.
Profilometer readings indicate the 1000' (308 m) long
control sections have smoother pavements than the fabric
sections. Soil boring revealed in the fabric sections
the bentonic clays were within 2.5' (0.77 m) of the
surface while they ranged from 5' to 10' (1.4 m to 3.1 m)
in the control areas. Initial moisture readings seemed
to show less change inside the DVFMB sections.

Another west Texas IH 10 geomembranes project was
part of a 48 mile (72 KM) asphaltic concrete resurfacing
in Culberson and Jeff Davis Counties completed in 1985.
The 40442' (12327 m) DVFMB had a trench gravel backfill
with an 18" (0.5 m) cement stabilized base cap. A
Phillips Petromat MB was used with a backhoe for
excavation, an average daily production of 400' (123 m),
and a one day high of 1000' (307 m). In 1991 a few
pavement distortions were reported. Some were in areas

where no DVFMB were placed and in others it was not placed due to construction conditions. The DVFMB sections in 1993 were doing fine.

Four contracts using geomembranes took place along IH 10 eastward from San Antonio toward Sequin. The first section from FM 1516 to Cibolo Creek, the Bexar-Guadalupe County line, constructed in the 1960's had a lime treated subgrade, foundation course, flexible base and a hot mix asphaltic concrete pavement (HMAC) as did the following three IH 10 sections. Expansive clays were beneath the treated subgrade. Smooth riding initially, the pavement soon began to distort and pavement maintenance costs $100,000 to $200,000 annually occurred on the 13 mile (19.5 m) section. The 1985 contract included 131200' (40370 m) of DVFMB bid at $13/LF (0.304 m). A trencher was used achieving a daily average of 900' (277 m) with a one day high of 2195' (675 m). Trench backfill was to have been limestone screenings with a cement stabilized cap. Screenings became difficult to secure, limestone scalpings, a finer material, was substituted. Shoulder cracking and depressions developed. Nondestructive testing revealed backfill voids. A contract was awarded to place a flowable portland cement, fly ash and sand slurry. In 1993 there's been some movement, some level ups, but overall the pavement serves well, and after the remedial work, so do the shoulders. Annual maintenance is described as nothing like it was before. (Personal Communication Stein 1/93)

The IH 10 approaches to Santa Clara Creek, developed pavement problems and 1987 rehabilitation work included 12000' (3692 m) of DVFMB bid at $15/LF (.304 m). A trencher was used, a Reemay (formerly DuPont Typar) placed and the project continues to hold up well in 1993. Extending westward from the IH 10 Santa Clara Creek project a rehabilitation contract using DVFMB to the Bexar Guadalupe County line, awarded in August 1988, included 14000' (4267 m) of Reemay fabric, bid at $11.42/LF (0.304 m) with a gravel trench backfill. Completed in 1989 its been holding up well with no problems. Another IH 10 contract extended east from Santa Clara Creek 4.3 mi (6.4 km) to Sequin with a segment on US 90. The work included 52800' (16246 m) of DVFMB bid at $11/LF (0.304 m). A Phillips fabric was used with a backfill of graded aggregate. Some caving occurred in areas of the existing cracked shoulders. The trench was moved further away from the shoulder which solved the problem. Since the August 1989 completion, the pavement has held up well. (Personal Communication Stein 1/93)

Sunraysia Highway Victoria Australia

VIC Roads constructed their first geomembrane test section in 1985. The ten year old highway section in the Wimmers desert region had developed longitudinal cracking with rutting up to 40 mm (1 1/2"). The 154 m (505') of fabric was placed vertically along both shoulders in trenches 1.5 m (4.9') deep and laid horizontally along the shoulders and roadway 5.5 m (15'). The trench excavated with a backhoe used a native clay backfill. An adjacent second section installed in early 1989, had geomembranes placed 2.5 m (8.2') and 2.8 m (9.1') deep. The conclusion was that placing the fabric at the deeper depths greatly reduced or virtually eliminated seasonal movements. (Holden, 2/92)

SH 112 Wyoming

In 1987 the Wyoming DOT used 251,900 SY (210,614 m²) of geomembrane vertically and horizontally. On a SH 112 7.8 Mi (11.7 Km) rehabilitation project the fabric bid at $0.50/SY (0.836 m²) was placed in trenches from 2.5' to 4' (0.7 to 1.2 m) deep and across the roadway. The horizontal section was covered with a 6" (150mm) sand blanket. Three other geomembrane projects were awarded that year, using horizontal and vertical placement. There were ten Wyoming geomembrane projects in 1988. One on SH 59 between Douglas and Gillette placed 77811 SY (65845 m²) horizontally and vertically, 2.5' to 4' (0.6 m to 1.3) deep. A 6" (250 cm) sand blanket was placed over the horizontal sections and native excavated soil was used as trench backfill. The bid price for the fabric was $1.44 SY (0.836 m²). The other projects with two on IH90 ranged downward in quantity to 180SY (144 m²). (Personal Correspondence Hager 1/93)

IH 25 Wyoming

In 1989 three WY DOT plans used geomembrane horizontally and vertically. Vertically the typical depth was 4' (1.2 m). On IH 25 between Kaycee and Buffalo 29822 SY (24931 m²) were used, bid at $1.80. Records for the other two projects with 55070 SY (46039 m²) of fabric are presently incomplete. Ten 1990 projects called for the same geomembrane installation. SH 315 had 101774 SY (85093 m²) bid at $1 per SY, extended from Mountain St. Lane to Decker Road. Two of the other projects US 212 and US 85 called for 233,000 SY and 210,210 SY (194,811 m and 175,757 m) each bid at $1.38. In 1991 Wyoming contracted five projects using geomembranes quantities from 218300 SY (182499 m²) to 200 SY (167 m²). Bid prices usually

averaged $1.38. The contract on US 85 between Mule Creek
Junction and Newcastle had 195,880 SY (163,756 m^2) of
geomembrane bid at $1.32.

A 1992 contract on SH59 between Douglas and Gillette
a 7 ml (11km) project included 230,000 SY (192289 m^2) of
fabric bid at $1.18. Six other sets of plans by the DOT
that year included geomembranes, five on roadways and one
for a ditch placement. In 1993 five geomembrane
contracts are ready for bids. Their quantities range
from 254,000 SY to 109,000 SY (212344 m^2 to 91124 m^2).
Between the years of 1982 and 1986 plans that included
227959 SY (190566 m2) of geomembrane were constructed.
The search for these records is continuing. Excavation
for the vertical placement was done with a Ditch Witch
machine. (Personal Communication Hager, 1/93)

IH 90 Montana

In the vicinity of the Montana Wyoming stateline a IH
90 rehabilitation project placed a geomembrane 3' to 4'
(0.9 to 1.2 m) deep along its shoulders. In the Big Horn
uplift its subgrade was undercut 2' to 3' (0.6 to 0.9 m)
deep and the fabric placed along several sections of the
23 mile (34.5 km) contract. The work was completed in
1986 and the pavement remains in fine shape. A
rehabilitation contract on IH 15 in the Great Falls area
had a 5' (1.5 m) excavation of subgrade which was
redensified, a geomembrane placed vertically and "maybe
horizontally". A Vermeer trencher was used to make the
cut. Difficulty in trenching deeper than 4' (1.2 m) was
experienced and was hard to follow up on projects.
Another IH 15 project is north of Great Falls, called for
subexcavating 2' (0.6 m) deep and replacing the expansive
material with sand. A geomembrane was placed vertically
along the shoulders. "There seems to be little interest
in dealing with the expansive soils". (Personal
Communication Yarger 11/92)

IH 40 Arizona

IH 40 a DOT rehabilitation contract completed in 1992
extended from the Apache County line to Pinta. The
geomembrane was placed vertically 7' (2.15 m) deep along
the shoulders with a 1' lap to the pavement, bid at $10
per foot (0.3 m). A rehabilitation contract on SH 88 took
place in the Globe area several years ago. The fabric was
placed vertically along the shoulders 7' to 8' (2.1 to
2.4 m) deep. To date no roadway deformations have
returned. Between Ash Creek and Sycamore Creek an SH 188
rehabilitation contract was awarded using a geomembrane

horizontally placed at a 3' (0.8 m) undercut and
vertically to a minimum depth of 7' (2.1 m). A second
project on SR88 in the Ash Creek area also used the
geomembrane similarly. Three other Arizona highways are
using or in the design phase considering the use of
geomembranes vertically. They are IH 40 east of Holbrook,
SR 95 in the Mojave City area and SR 666 in the Lutresso
section. (Personal Communication Beekman 1/93)

IH 45 Texas

     This major link between Dallas and Houston, IH 45
near Waxahachie had a 16.4 mi (24.5 km) 1990 contract to
improve the concrete pavement distorted by expansive
clays. It included 285,140' (87735 m) of DVFMB bid at
$6/LF (0.308 m). Early problems with trench caving were
solved by reducing the delay in placing the coarse sand
backfill. Daily placement rates ranged from 1500' to
1800' (462 m to 554 m) with a one day high of 2400' (740
m). The project continues under construction. (Personal
Communication Stanford, 2/93)

     Fifteen other projects have been identified using
DVFMB. Eleven are highways, one each in Alabama,
Kentucky, California and Georgia, with an additional
seven in Texas. Quantities ranged from 46000 SY
(368000m^2) to 700 SY (560 m^2). (Table IA) Four other
projects were airfields, two each in California, and
Columbus, SA.

Conclusions

     There are many vertical and horizontal barriers used
in an effort to control the destructive movements of
expansive soils on transportation facilities. Over 100
projects have been identified as using geomembrane
barriers. These water-proofed barriers are meeting the
challenges of expansive soils that have been identified
in 45 of the 50 United States as well as all the world's
continents excepting those in the polar regions. The
information collected to date indicates that many
agencies use the geomembranes but often are unable to
follow a post construction testing pattern due to lack of
funding and personnel.

     Results available clearly indicate that the
geomembranes can be placed vertically. Backfill material
must be addressed with care as use of unconsolidated fine
material can cause subsequent subsidence. Whether
horizontal or vertical placement is more effective or a
combination of both is needed has not been clearly

defined. Results do clearly indicate that the geomembrane barriers generally reduce the destructive movements of the expansive soils. The cost of placing the fabric vertically has come down considerably, and the savings of using the geomembranes has increased significantly. The geomembranes have been used widely and are effective.

References

Browning, G. "Evaluation of Soil Moisture Barrier Construction", Report No. 61-21, 11/92, Mississippi Department of Transportation.

Holden, J.C. "Reduction of Pavement Damage From Expansive Soils Using Moisture Barriers", 2/92.

McDonald, E.B. "Experimental Moisture Barrier & Waterproof Surface", HR 0200 (364), South Dakota Department of Transportation, 10/73.

Spensko, A. and Chichester, "Fabric Interlayers to Prevent Defective Cracking on Asphalt Pavements", Experimental Project UT 81-04 UTAH Department of Transportation, 4/87.

Steinberg, M., "Deep Vertical Fabric Moisture Seal." Fourth International Conference on Expansive Soils, pgs. 383-400 3/80.

Steinberg, M., "Horizontal Placement of a Geotextile on a Subgrade to Control a Swelling Soil", 187-9 Texas DHDT, 2/83.

Steinberg, M., "Controlling Expansive Soil Destructiveness by Deep Vertical Geomembranes on Four Highways," pgs 48-53, TRB Record 1032, 1985.

Steinberg, M., "Controlling Expansive Soils: Twenty Texas Highway Projects", Seventh International Conference on Expansive Soils, pgs. 392-397, 8/92.

Wu, X.M., Wei, D.X., c/ Chen, J.H., "Damage to Road Beds & Remedial Measures", Seventh International Conference on Expansive Soils, 8/92.

TABLE I - A Barrier Sampling						
Place	Highway (or RR)	Quantity	Bid Price $	Horiz or Vert	Year	Perfor mance
Texas	MMullen	600'	--	H	1976	Good
Texas	IH 410 (S.A)	4978'	$20.00	V	1978	Good
Wyo	IH 80	108,854'	--	H	1981	Good
Texas	IH 37 (S.A.)	21,483'	$21.00	V	1981	Good
Texas	IH 10 East SA	131,200'	$13.00	V	1985	Good
Astrla	Sundray sia	600 m	--	V	1985 1989	Deeper Better
China	Shanxi R.R.	600 m	--	H	1989	Good
Miss	IH 20	142,820 SY	$2.35	H	1990	Too Soon
Texas	IH 45	285,145'	$6.00	V	1990	✓✓✓✓
Wyo	SH 59	230,000 SY	$1.18	H/V	1992	✓✓✓
TABLE IA						
Texas	IH 30	4000'	$13.58	V	1984	?
Texas	SH 97	5600*	$10.00	V	1986	poor
Texas	FM 465	3100'	$10.00	V	1986	poor
Texas	FM 725	6840'	$20.00	V	1986	poor
Cal	IH 80/ US 50	2500'	---	H-V	1987	good
Kentky	IH 24	5 @ 500'	---	H	1989	good
Texas	FM 1516	14400'	$10.00	V	1989	good
Texas	US 84	46,000'	$10.50	V	1990	good
Texas	IH 635	3836/700	---	H-V	1989 1993	?

DATA BASE FOR THE DESIGN OF PIERS IN EXPANSIVE SOILS

M. Picornell[1], C. Ferregut[1], H. Gowdappa[2], and J. Behar[2].

Abstract

The paper describes the parameters and the structure of the seed of a data base of properties and observed field behavior of expansive soils assembled from an extensive literature review. The data base has been designed in such a manner that it could easily be integrated into reliability analysis programs or used in conjunction with expert systems for the analysis and design of piers in expansive soils. The paper also describes the variability of the compiled data from several statistical analyses of the more influential parameters assembled in the data base.

Introduction

Drilled shafts are a common type of foundation for heavy structures in areas that have problems with expansive soils. The piers are usually designed to transmit the structural loads to the soils below the surficial active zone. Furthermore, the piers have to be designed to withstand the tensile stresses that the swelling of the soil would induce in the pier. A deterministic design of the pier against these tensile stresses would require some site specific information that is too costly and too time consuming to be developed for each specific site.

Previous work in this area (Ferregut and Picornell, 1991) has suggested that the site specific information needed in the design of the piers has quite different levels of importance in defining the stresses in the pier. Consistent with this finding, a reasonable approach could be that the more influential parameters are developed for each specific site while the less influential parameters are selected from previous experience elsewhere in the area, the state, or the world.

----

[1]Associate and Assistant Professors, and
[2]Graduate Research Assistants, Center for Geotechnical and Highway Materials Research, The University of Texas at El Paso, El Paso, Texas 79968-0516

These considerations suggest the need of a data base of soil properties and observed soil field behavior of expansive soil deposits that could be drawn upon to complement site specific information. The present article describes the parameters and structure of a proposed data base designed to easily interface with reliability analysis methodologies, such as the one proposed by Ferregut et al.,(1992) or to be easily drawn upon by expert systems for the design of drilled piers.

Data Base Description

The information used to compile this data base was extracted mainly from the Proceedings of the seven International Conferences on Expansive Soils that have been held since 1969 around the world. All the Proceedings of these conferences were reviewed having in mind two main criteria to select an article as a potential source of information. These two criteria were that the article presented field measurements of swelling versus depth and/or surface heave versus time. When an article was found with some of these data, the article was selected and subjected to further review. The majority of the data were found in graphs, and sometimes each graph contained more than one observation. Each one of these observations was treated as a different case and it was assigned a specific case number into the data base. In order to obtain the values that shape each curve it was necessary to use a digitizer, since as mentioned above, most of the times the data were given in graphs and very seldom in table form.

After all the information that was requested by the data base format was identified within the article and the graphical data was digitized, we proceeded to fill out the data base format which consisted of three pages of data for each individual case. For illustration purposes, the data of Case No. 10 is documented below in Figs. 1 and 2, this case was taken from the 4th International Conference on Expansive Soils.

Figure 1, shows the first page of data collected for this case. Information is gathered about the author and publication of the paper, as well as the type of structure the soil withstands and some characteristics of the soil layers. Figure 2 shows the data on surface heave versus time and swelling versus depth. The columns labelled S1 through S5 are reserved to record heave versus depth at five different times, and there is space at the bottom of the page for recording these dates. Sometimes, not all the information was available and thus some spaces in the data formats were left blank. A third set of information recorded includes any published data on the environmental conditions during the period of field measurements. Specifically, the monthly Potential Evapotranspiration and monthly rainfall depth are included in the data base when published.

The data of heave versus depth and surface heave versus time was first normalized and then entered in the data base. The procedure to normalize this data has been explained in detailed elsewhere (Ferregut et. al., 1992). In order

CASE No. 0 0 1 0

AUTHORS (Last, Initial) Johnson, L.D.

TITLE Field Test Sections on Expansive Soil.

DATE June 1980   **PUBLICATION** 4th International Conference on Expansive Soils. Volume I. pp. 262 - 283.

GEOGRAPHICAL LOCATION:

Clinton,   Miss.   USA.
CITY   STATE   COUNTRY

TYPE OF STRUCTURE:

BARE	**VEGETATED**	**SLAB**			**ROAD**	
		EDGE	**CORNER**	**CENTER**	**CENTERLINE**	**EDGE**
☐	☐	☑	☐	☐	☐	☐

SUBSURFACE CONDITIONS:

UPPER LIMIT (ft)	LOWER LIMIT (ft)	CHARACTERISTICS (Soil Type, Atterberg Limits, water content(w), Swelling Pr.,etc.)
0.0	7.85	Loess overburden brown, CL, $w = 22\%$
7.85	24.89	Weathered Yazoo tan CH, $PL = 33\%$, $LL = 98\%$, $w = 48\%$
24.89	50.0	Unweathered Yazoo Clay gray, CH.

WATER DEPTH BELOW SOIL SURFACE 5.92 **ft.** (Perched Water Table)

FIGURE 1. First page of data format.

SWELLING DATA

SURFACE HEAVE vs TIME		HEAVE vs DEPTH					
s (in) vs	t (months)	Heave (in)					Depth (in).
		s1 @T1	s2 @T2	s3 @T3	s4 @T4	s5 @T5	
-0.1169	4.90						
-0.1615	8.49	-0.1974	0.3834	0.747	1.1150	1.319	0.0
-0.1145	15.07	-0.1746	0.3712	0.6548	1.0650	1.220	60.0
-0.0018	16.94	-0.1495	0.3140	0.6144	0.9700	1.130	120.0
0.0511	16.94	-0.0968	0.2755	0.4980	0.7840	0.900	240.0
0.0795	19.94	-0.5150	0.1805	0.1880	0.6000	0.663	360.0
0.1656	23.09	0.0000	0.0000	0.1220	0.345	0.350	420.0
0.2373	25.98			0.0000	0.120	0.125	485.0
0.4752	41.65				0.000	0.000	520.0
0.5109	44.57						
0.5989	48.22						
0.7371	54.94						
0.7695	61.63						
0.9188	68.30						
0.9370	73.46						
1.0247	80.13						
1.0460	88.68						
1.1472	93.38						
1.2271	102.21						
1.3123	118.70						

STARTING RECORD YEAR OCT/1969   T1 = 6/1970 **T2 =** 6/1971 **T3 =** 1/1975 **T4 =** 7/1977 **T5 =** 6/1979
ENDING RECORD YEAR 1979

FIGURE 2. Second page of data format.

to normalize the data, estimates of the depth of the active zone and the maximum surface heave were necessary. When this data was not recorded in the original publication, the data was extrapolated from the published measurements. In this sense, the most common case was when all the points of observations experienced some heave. In that instance, the depth of active zone was selected by extrapolating to the depth where zero heave was expected based on the trend of heave versus depth indicated by the published measurements.

Data Base Manager

An expert system for the analysis and design of piers in expansive soils is currently being developed by the authors. In general an expert system is a computer program that attempts to capture the knowledge and experience of one or more human experts in order to make such expertise available on demand to the user of the program. The commercial expert system development shell EXSYS Pro (EXSYS, 1992) is being used to develop the expert system. As part of the expert system a database manager module has been developed to conduct basic statistical, and reliability oriented data analysis. This module provides a useful set of illustrative menu driven tools to manipulate and sort data, perform basic descriptive statistics, curve fitting and parameter estimation. Some of these tasks are conducted by linking EXSYS Pro to other commercial software. Figure 3 shows the relationships among all the component modules of the expert system. The main components of the database are the swelling versus depth and the swelling versus time data. These data are stored in separate ASCII files for each case identified in the technical literature. The data is grouped according to the country and every country in the database is given a country code number for reference. Within each country, the records are categorized according to the state and location of origin. Every data within a country is given an identification case number. The code numbers facilitate the viewing, sorting and merging of the separate data cases. The inherent capabilities of EXSYS Pro are used to perform these tasks.

A key step in the analysis of piers in expansive soils is the analytical modeling of the heave versus depth and heave versus time curves. Once a particular data set is chosen, this modeling can be done using the curve fitting capability of the expert system. This capability consists of linking a commercial software TABLECURVE (Jandel Scientific, 1992) which allows the user to fit a library of up to 3000 analytical functions to the data. In addition the user can specify his/her own equations. To use the curve fitting software the user is provided with help screens which guide the user through the software. Figure 4 and Figure 5 show curves fitted to swelling versus depth and swelling versus time data respectively.

In order to perform a reliability analysis of piers in expansive soils, the user needs to know the statistical distributions of a number of parameters. To

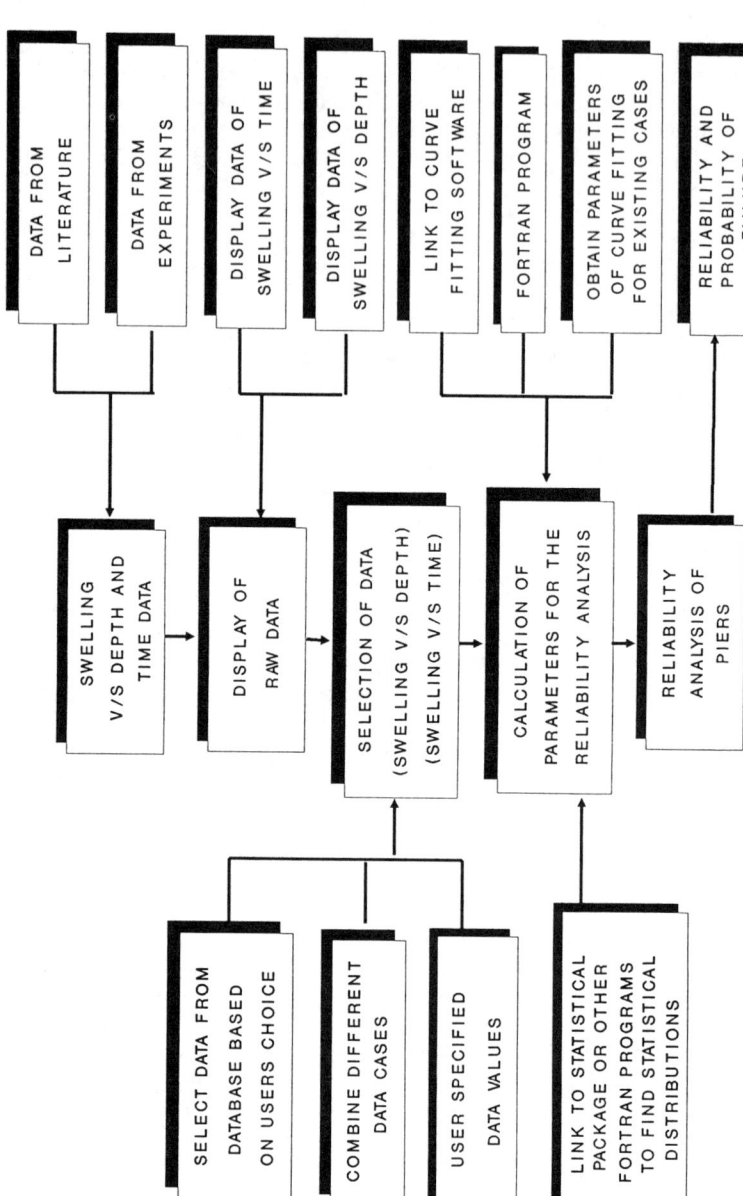

Figure 3. Relationship Among Different Modules of Expert System

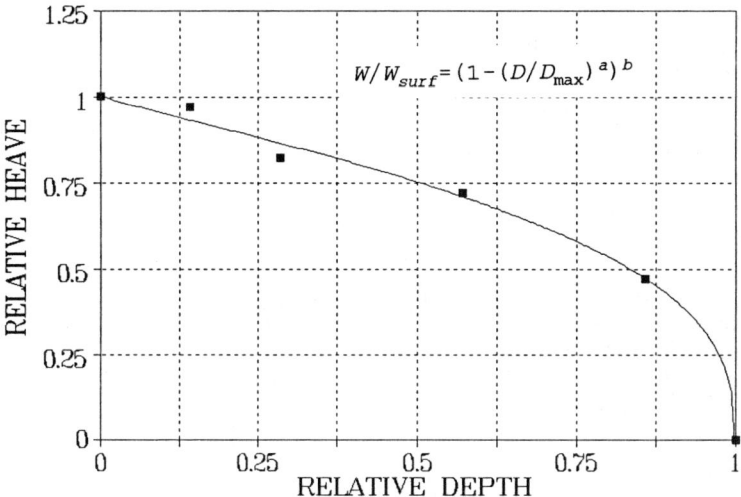

Figure 4. Curve Fit to Relative Heave Versus Relative Depth Data

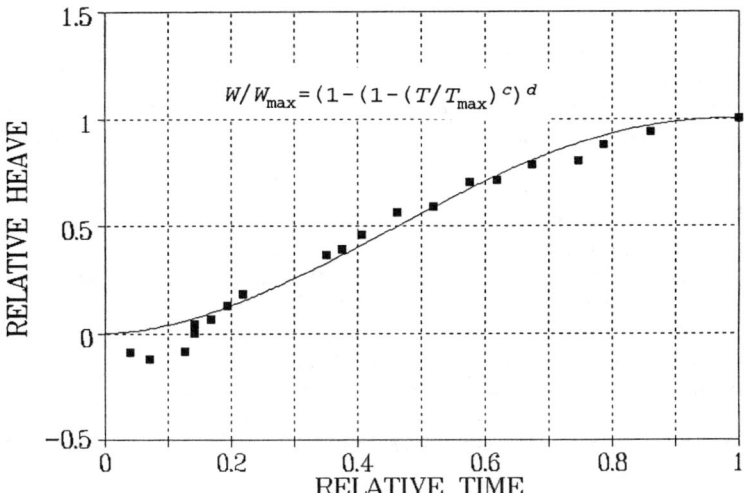

Figure 5. Curve Fit to Relative Heave Versus Relative Time Data

conduct this task the user can use the statistical analysis package or the
FORTRAN programs which are linked to EXSYS Pro. Figure 6 shows a fitted
distribution to one of the parameters which influence the reliability of piers in
expansive soils. Once the statistical distributions for the key parameters are
known, the reliability analysis of the pier is conducted using First Order
Reliability Methods (Ferregut and Picornell, 1991). A FORTRAN subroutine
is linked to EXSYS Pro to perform this task.

Variability of Soil Parameters

Heave Profile. A preliminary study (Ferregut and Picornell, 1991)
identified the heave profile occurring in the soil around the pier as one of the
most influential factors on the tensile stresses and displacements experienced
by a pier in expansive soils. The authors proposed the following expression to
model the dimensionless heave profile,

$$\frac{W}{W_{Surf}} = (1 - (\frac{D}{D_{max}})^a)^b \tag{1}$$

Where, W is the heave at depth D; W_{Surf} is the surface heave; D_{max} is the depth
of the active zone; and a and b are parameters. Uncertainties on the heave
profile are taken into account by considering all parameters in Eq. (1) to be
random variables.

A total of 45 sets of heave vs depth data currently exist in the data base.
Using these data sets, statistical estimates and probability distributions were
computed and fitted to the variables in Eq. (1) and parameters a and b.
Results for parameters a and b are shown in Figs. 6 and 7. Several probability
distributions were fitted to the histograms of these two parameters.It was
observed that both parameters can best be modelled using a Lognormal
distribution. This agrees with the results of Ferregut et al (1992) conducted
with a more limited data set.

Surface Heave.The data base currently has 36 surface heave vs time
curves. To facilitate the comparison of the different sets of data,
Pendones(1992) proposed that the profiles be described in terms of
dimensionless relative surface heave and relative time. This is accomplished
by dividing the time into the time required to reach the maximum swelling and
the surface heave into the maximum surface heave recorded. The expression
proposed by Pendones to model the resulting normalized curves takes the form:

$$\frac{W}{W_{MAX}} = 1 - (1 - (\frac{T}{T_{MAX}})^c)^d \tag{2}$$

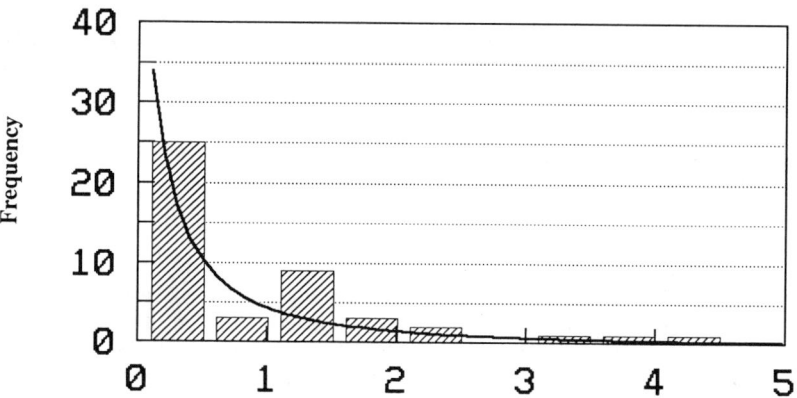

Figure 6. Histogram and Best Fit Distribution for Parameter - a

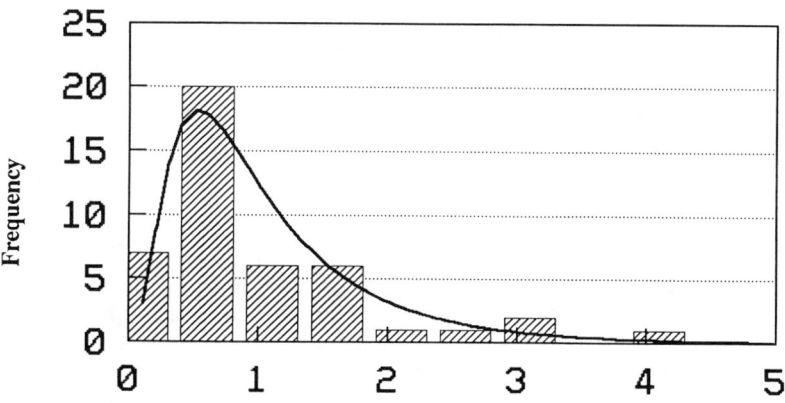

Figure 7. Histogram and Best Fit Distribution for Parameter - b

where: W is the heave at time T, W_{max} is the maximum surface heave, T_{max} is the time at maximum heave, and c and d are parameters. Best fit distributions to the histograms of parameters c and d are shown in Figs. 8 and 9. The best fit distribution for parameter c is an exponential distribution and for parameter d is a lognormal distribution. Unlike the previous case, this results differ from the results of Ferregut et. al., (1992) who identified the Gamma distribution as the best fit for both of these parameters. However, their results were based on a more limited data base.

Practical Application

The data base can be integrated into a reliability based design at several different levels. In this manner, it is possible to design the pier based exclusively on the data base and without any site specific information. Nevertheless, if the user has developed some site specific information, this can be complemented with other information from the data base.

In this sense, some site specific information such as the surface heave or the depth of the active zone might be available from previous experience in the area or can be reasonably be estimated for the site. By way of contrast, other data such as the time of maximum swelling and the parameters that describe swelling versus depth and swelling rates are not as likely to be available from previous experiences. Furthermore, to develop estimates of these parameters for a site would be quite involved.

The user would have to specify the expected value of the site specific information and the corresponding coefficient of variation. For the data from the data base the user can select several possible alternatives such as the following: (1) distributions have been already fitted to all existing cases in the data base and can be readily incorporated in the analysis, or (2) the existing cases in the data base from a particular region, state, or nation can be selected to form histograms of the corresponding parameters and fit new probability density distributions.

Conclusions

The initial concept design of a data base of swelling soil parameters has been proposed. The data base has been designed with a data manager that facilitates its interface with reliability analysis and design programs and with expert systems dealing with drilled pier design. Even though the capabilities and current data in the data base are limited to a specific application, it can easily be expanded to incorporate other information that may be needed for other types of foundations.

The main advantages offered by the data base are the following: (1) easy updating as information becomes available; (2) it allows to correlate swelling behavior with soil properties, environmental conditions and geographical

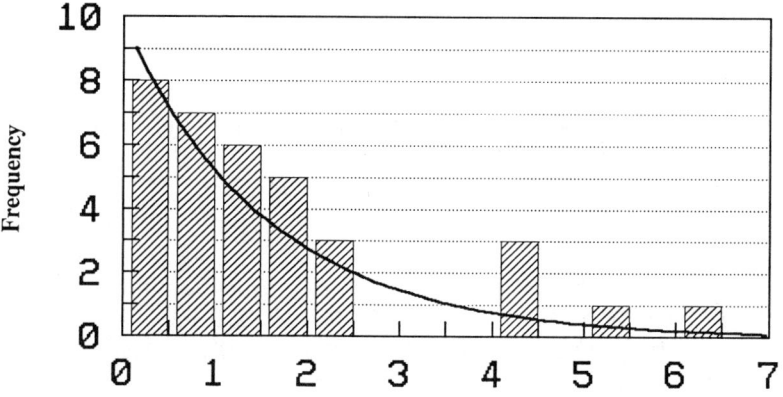

Figure 8. Histogram and Best Fit Distribution for Parameter - c

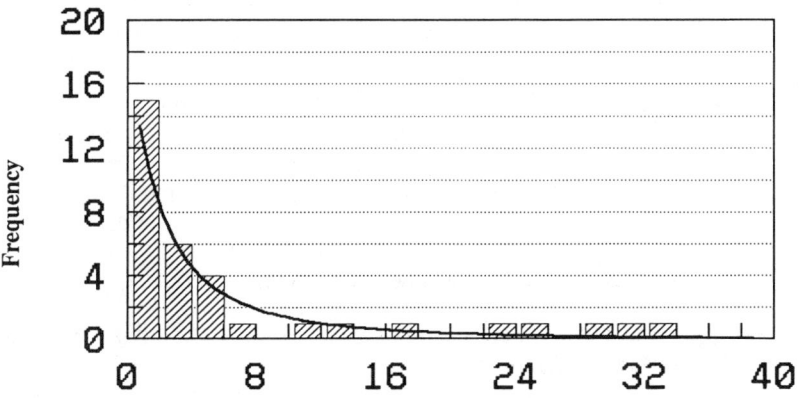

Figure 9. Histogram and Best Fit Distribution for Parameter - d

region; and (3) it facilitates the development of statistical and probabilistic model of soil behavior.

Acknowledgements

The work reported has been performed as a part of Project Number 3661-022 funded by the "Advanced Research Program" of the Texas Higher Education Coordinating Board. This support is gratefully acknowledge.

Bibliography

EXSYS, 1992, "EXSYS Professional for Microsoft Windows," Manual for an Expert System Development Software, EXSYS, Albuquerque, New Mexico.

Ferregut, C., and M. Picornell, 1991, "Reliability Analysis of Drilled Piers in Expansive Soils," Canadian Geotechnical Journal, Vol.28, No.6, pp. 842.

Ferregut, C., M. Picornell, and J.A. Pendones, 1992, "Probability Analysis of Piers in Expansive Soils," Transportation Research Record, No.1362, Transportation Research Board, National Research Council, pp. 118-125.

Ferregut, C., M. Picornell, and J.A. Pendones, 1992, "Analysis of Uncertainties Related to the Response of Piers in Expansive Soils," Proceedings of the 7th International Conference on Expansive Soils, Dallas, Texas, pp. 477-482.

International Conference on Expansive Soils,
		Proceedings,		1st - College Station, Texas, USA. (1965)
						2nd - College Station, Texas, USA. (1969)
						3rd - Haifa, Israel. (1973)
						4th - Denver, Colorado, USA. (1980)
						5th - Adelaide, South Australia. (1984)
						6th - New Delhi, India. (1987)
						7th - Dallas, Texas, USA. (1992)

Jandell Scientific, 1992, "Table Curve Manual," Version 3.12, AISN Software, Corte Madera, California 94925.

SUBJECT INDEX
Page number refers to first page of paper

AUTHOR INDEX
Page number refers to first page of paper